Special Topics in

FOUNDATIONS

Proceedings of a session sponsored by
the Geotechnical Engineering Division of the
American Society of Civil Engineers
in conjunction with the ASCE National Convention,
Nashville, Tennessee

May 9, 1988

Edited by Braja M. Das

Geotechnical Special Publication No. 16

ASCE 1852

Published by the
American Society of Civil Engineers
345 East 47th Street
New York, New York 10017-2398

ABSTRACT

This book contains papers presented at a session related to various theoretical and experimental developments on foundation analysis, design and construction. The papers were written by both researchers and practitioners. Topics include the development of a nonlinear model for prediction of movement of drilled shafts in expansive clay; new techniques for lime stabilization of expansive clay for foundation construction; evaluation of soil improvement achieved by sand columns for foundation construction by means of place load tests; uplift capacity of vertical multihelix, and inclined plate anchors; bearing capacity and settlement characteristics of shallow foundations on granular trench made in clay, and with geoweb used as reinforcement in granular soil.

Library of Congress Cataloging-in-Publication Data

Special topics in foundations: proceedings of a session sponsored by the Geotechnical Engineering Division of the American Society of Civil Engineers in conjunction with the ASCE National Convention, Nashville, Tennessee, May 9, 1988; edited by Braja M. Das.

p. cm.

Includes indexes.

ISBN 0-87262-645-8

1. Foundations—Congresses. I. Das, Braja M., 1941- II. American Society of Civil Engineers. Geotechnical Engineering Division. III. ASCE National Convention (1988: Nashville, Tenn.)

TA775.S65 1988 88-6507

624.1'5—dc19 CIP

Library of Congress Catalog Card No.: 88-6507
ISBN 0-87262-645-8
Manufactured in the United States of America.

PREFACE

This special technical publication constitutes the proceedings of a session, the objective of which is to provide and update the theoretical and experimental developments in the analysis, design, and construction of foundations. Of the twelve abstracts initially proposed, seven papers from the U.S.A., Canada, and Belgium were finally accepted. These papers were written by academic researchers and field practitioners. The papers cover a wide variety of topics, such as the development of a nonlinear model for prediction of movement of drilled shafts in expansive clay; new techniques for lime stabilization of expansive clay for foundation construction; evaluation of soil improvement achieved by sand columns for foundation construction by means of plate load tests; uplift capacity of vertical multihelix and inclined plate anchors; bearing capacity and settlement characteristics of shallow foundations on granular trench made in clay, and also with geoweb used as reinforcement in granular soil.

It is the current practice of the Geotechnical Engineering Division that each paper published in a special publication be reviewed for its contents and qualities. Reviews of papers for publication in this volume were conducted by the Geotechnical Engineering Division's Committee on Shallow Foundations in coordination with the Division Committee on Publications. The following individuals reviewed the papers:

J. L. Briaud
S. P. Clemence
B. M. Das
J. R. Davie
T. B. Edil
M. W. O'Neill
V. K. Puri
R. W. Stephenson
W. K. Wray

Each paper included in this volume has received two positive peer reviews. All papers published are eligible for discussion in the *Journal of Geotechnical Engineering* Division, and they are eligible for ASCE awards.

Thanks are extended to the authors of the papers and to the reviewers for their timely response. I am grateful to Janice Das for typing all correspondence relating to this session and making some last minute adjustments. Thanks are also due to Shiela Menaker for her support throughout the publication process.

Braja M. Das

TABLE OF CONTENTS

ADAPTIVE MODEL FOR DRILLED SHAFTS IN EXPANSIVE CLAY

Michael W. O'Neill,[1] M. ASCE

ABSTRACT

Methods for predicting free-field movements in clay soils subjected to changes in suction are applied to the prediction of "translating" unit soil resistance-movement curves for the shaft and base of a drilled shaft or footing. These curves are formulated in terms of hyperbolic tangent functions that are offset with respect to the zero-displacement/zero-resistance origin by an amount equal to the computed free-field swell displacement. Consideration is given to the prediction of peak shear stresses for these curves through the use of zero swell pressures on laterally oriented clay specimens.

An efficient nonlinear model is described whereby a flexible drilled shaft or footing, displacement-dependent imposed loads, and soil anchorage are modelled to provide predictions of shaft movement and stresses in the plinth.

INTRODUCTION

Drilled shaft foundations are often used to bypass surficial soils that have a high shrink-swell potential but that would otherwise possess strength and compressibility properties adequate to support shallow foundations.

Often, drilled shafts can be designed by assuming the shaft to be rigid and immovable. However, where shafts are flexible and the zone of soil expansion extends to a significant depth, a soil-structure interaction analysis may be of value during the design process.

LOAD TRANSFER IN EXPANSIVE CLAYS

Movement of drilled shafts in expansive clay can often be reduced by providing anchorage in moisture-stable soils (or rock) below the zone of seasonal moisture change by extending the shaft to an adequate depth when such soils are not moisture deficient. Anchorage against uplift loads imposed on the drilled shaft by expanding soils near the surface is provided by a bell and/or skin friction.

Fig. 1 illustrates the anchorage requirement. In expansive soil, under the same load Q_T as in non-expansive soil, the upward-directed shear stresses f along the shaft can be much higher than those in the non-expansive soil because the soil has moved upward with respect to the shaft in addition to the shaft's having moved downward (under compression loading) with respect to the soil. The

---

[1]Prof. of Civil Engineering, University of Houston, Houston, Texas 77004.

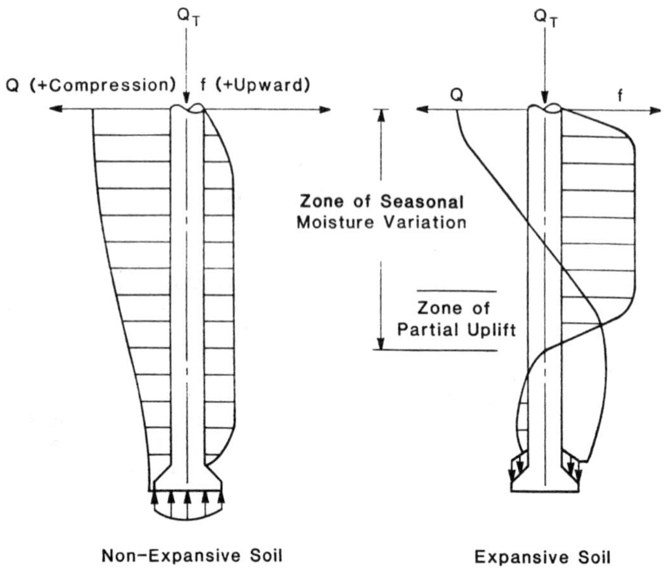

Fig. 1. Idealized Loads and Shaft-Soil Interface Stresses at Working Load

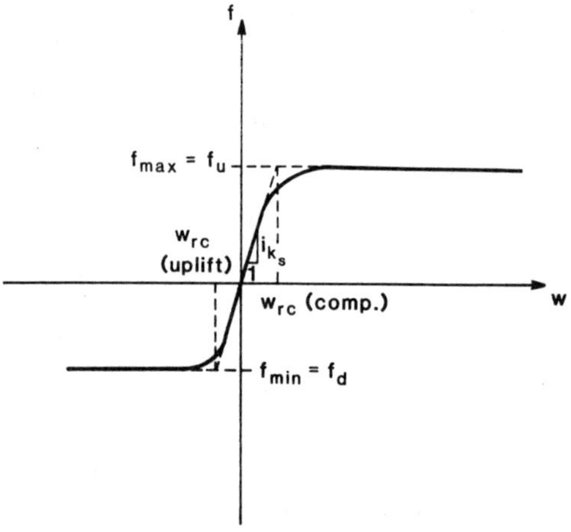

Fig. 2. Hypothetical Unit Load Transfer Vs. Movement Relationship

soil may be viewed as having a fundamental unit side shear (f) - displacement (w) relation at any depth that is uncoupled from similar relations at other depths, as shown in Fig. 2. If the soil is also viewed as being perfectly plastic after failure, then a limiting unit side shear is reached at some critical relative deflection, w_{rc}, defined in Fig. 2, where w_{rc} results from the sum of the upward movement of the soil and downward (or upward) movement of the shaft.

The magnitude of the upward-directed limiting value of f (= f_u) is considered independent of Q_T in expansive soils. Through the zone of seasonal moisture variation, f_u is obviously a time-dependent variable, making total stress methods difficult to apply to its evaluation. As the soil swells and moves upward relative to the shaft, its strength tends to change. Effective stress methods can be applied to the evaluation of f_u. As the soil imbibes water, it swells not only vertically but laterally until it is in contact with the face of the shaft (Fig. 3). Beyond this point the soil swells with zero radial strain at the interface. Although the soil may have undergone shrinkage between the time of construction and the design swelling event, it can be assumed approximately that the process of lateral soil shrinking and swelling back to the surface of the shaft is reversible, so that the normal effective pressure placed on the face of the shaft during swelling is equivalent to the zero swell pressure of a horizontally oriented sample of soil (σ'_{os}) with an initial suction equal to that at the time of shaft construction. Assuming that the drilling operation and expansion produced shearing produces enough destructuring in the soil at the interface that residual conditions are achieved and that loading is slow enough to prohibit accumulation of excess pore water pressures,

$$f_u = \sigma'_{os} \tan \phi'_r, \qquad (1)$$

where ϕ'_r is the residual angle of internal friction of the soil. If loading is rapid, pore water suction may be possible due to soil dilatency, and drilling that produces a "rifled" borehole can reduce the presence of destructured soil on the failure plane and also increase the dilatency component of shearing resistance. These factors both make f_u larger than would be predicted by Eq. 1. On the other hand, σ'_{os} would normally be obtained in an oedometer in which zero lateral strain is permitted, making σ'_{os} larger than would occur in the field, where "lateral" (vertical for a horizonatally oriented element) strain is not zero. Therefore, Eq. 1 is modified to

$$f_u = a\ \sigma'_{os} \tan \phi'_r, \qquad (2)$$

where a is an empirical factor, greater than 1.0.

It is difficult to evaluate factor a through physical testing. One test was conducted by the US Army Corps of Engineers (2,6) in a clay-shale formation in San Antonio, Texas. This highly preconsolidated formation was overlain by clay, as shown in Fig. 4. In order to induce swelling, the test site was ponded, and a level of design load was placed on the shaft for approximately one year. Unit side resistance actually decreased in Zone A (overburden), increased in Zones B (overburden) and C (top of shale), and became negative in Zone D, where anchorage developed. However, f_u in the clay shale was

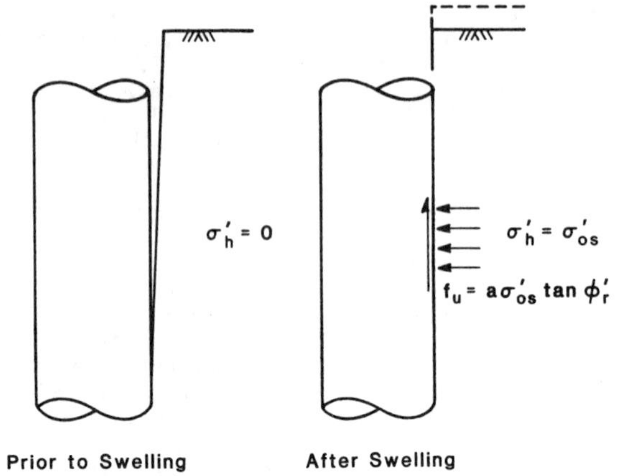

Fig. 3. Hypothetical Description of Effective Stresses Against Drilled Shaft Wall During Swelling

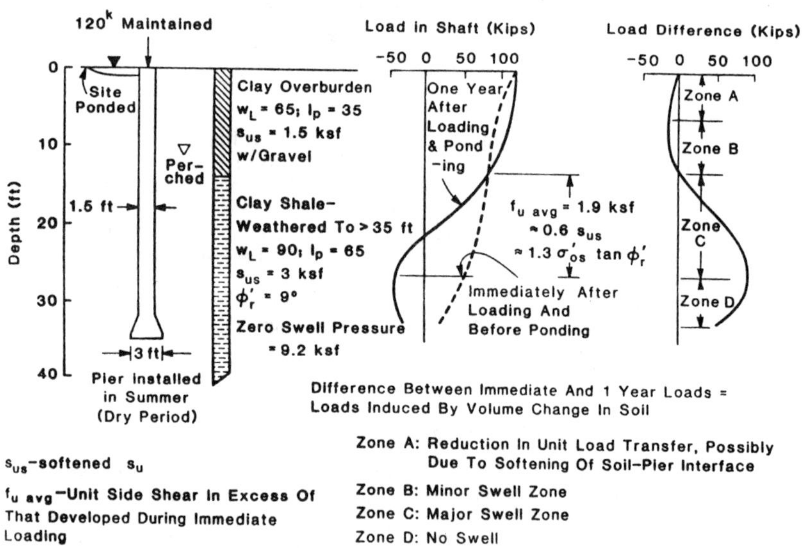

Fig. 4. Measured Relationship Between Unit Side Shear in Uplift and Cohesive Shear Strength, Zero Swell Pressure and Residual Angle of Internal Friction (6)

approximately 1.3 σ'_{ro} tan ϕ'_r (a = 1.3) on the average, or 1.9 ksf. Other studies of this nature are needed to establish design rules for assessing f_u.

For simplified design purposes, the shaft may be assumed to be rigid and immovable, and f_u is assumed to be applied over the entire depth of seasonal moisture change (evaluated later), except within the "zone of partial uplift," located at the bottom of the zone of seasonal moisture change. Here, the magnitude of the relative soil-to-pile movement is less than w_{rc}, so the limiting unit shaft resistance cannot be developed. It can be assumed that the upward-directed unit shaft resistance varies from zero at the bottom of the zone of seasonal moisture change to f_u at the top of the zone of partial uplift. The location of the top of the zone of partial uplift can be established as the depth at which the estimated free-field soil heave equals w_{rc}. This statement implies that a vertical free-field soil heave profile must be established, w_{rc} must be estimated and the depth z_d of the zone of seasonal moisture content must be obtained. Below the zone of seasonal moisture change, the shaft resistance is directed downward on the shaft (at least for relatively small values of Q_T), but if a bell is used as an anchor, shaft resistance in that area may be neglected for design purposes because of bell-shaft interaction effects. If a bell is provided as an anchor, the soil reaction is downward on the shoulder of the bell for small values of Q_T.

Estimation of w_{rc}

Given the assumption that f_u can be estimated, w_{rc} can be approximated from an expression for computing relative deformations of a cylinder and an elastic soil (7):

$$w_{rc} = \frac{f_u}{G_s} \{r_s \, [0.67 + \ln \frac{\zeta \ell}{r_s}]\} , \tag{3}$$

where r_s = radius of the shaft; ℓ = shaft length; G_s = shear modulus of the soil at low strain levels; and

$$\zeta = \frac{G_s \text{ at depth } \ell/2}{G_s \text{ at depth } \ell} \tag{3a}$$

Rigorously, G_s must be idealizable as varying linearly with depth. At any given depth w_{rc} can then be estimated using G_s at that depth.

Estimation of z_d

The estimation of the depth of seasonal moisture change may be done by several means, including long-term soil suction measurements, establishment of depths of ancient surface cracks, and establishment of depth of grass root zones. One method involves graphing liquidity index or water content/plastic limit vs. depth over several seasons (6). Envelopes to the discrete data tend to become constant with depth below z_d.

ESTIMATION OF FREE-FIELD HEAVE PROFILE

In many drilled shaft applications in expansive clays the base of the shaft is located above a ground water table. According to Lytton (3), the value of stable soil suction at depth z_d can be expressed as a function of a climate index factor, termed the Thornthwaite Moisture Index (TMI). General contours of the TMI in the United States are shown in Fig. 5. Site-specific calculations can be done according to Thornthwaite (11). Stable suction, according to Lytton, can be expressed approximately by Eqs. 4 and 4a:

$$\tau_o \text{ (in pF units)} = 5 - 0.05\,(\text{TMI} + 52) \quad (-52 \le \text{TMI} \le -25) \tag{4}$$

$$= 3.65 - 0.007\,(\text{TMI} + 25) \quad (-25 \le \text{TMI} \le 52) \tag{4a}$$

Initial ("dry") and final ("wet") soil suction profiles can be assessed for worst-case analysis as shown in Fig. 6. τ_{ws}, the wet season suction at the soil surface, is given by

$$\tau_{ws}(\text{cm}) = \tau_o(\text{cm}) - z_d(\text{cm}). \tag{5}$$

τ_{ds}, the dry season suction at the soil surface in cm of water, is obtained from site specific measurements (e.g., 4), but an approximation in many areas according to Lytton can be obtained by

$$\tau_{ds}(\text{pF}) = \tau_{ws}(\text{pF}) + 2. \tag{6}$$

Where anomalies exist (trees, slopes, etc), Eq. 6 is not valid, and site specific measurements would be required.

In order to calculate a free-field heave profile, the active zone between $z = 0$ and $z = z_d$ can be subdivided several times. At each depthwise subdivision

$$\left(\frac{\Delta H}{H}\right)_i = f\left[-\xi_\tau \log_{10} \frac{\tau_{final}}{\tau_{initial}} - C(\sigma)\right]. \tag{7}$$

τ_{final} is a value on line bc in Fig. 6, while $\tau_{initial}$ is a value on line ac. Both are obtained at the depth of the center of sublayer i. ΔH is the upward vertical movement in the sublayer, and H is the thickness of the sublayer. f is a factor to convert volumetric swell to vertical swell. Its value depends on the crack pattern in the soil. If the soil is highly cracked, so that the cracks cannot close fully in the wet season, f = 0.33. If the soil is uncracked in the dry season, f = 1.00. For most applications during swelling, $0.7 \le f \le 0.9$, so f = 0.8 may be assumed for ordinary design purposes. ξ_τ is the coefficient that relates swell movement to changes in suction pressures in the soil.

$C(\sigma)$ is a correction term for surcharge, which may be evaluated from

$$C\,(\sigma) = \xi_\sigma \log_{10} \left(\frac{\sigma_s}{\sigma_1}\right), \tag{8}$$

where σ_s is the total octahedral pressure at the depth of interest (external stress plus soil body stress) and σ_1 is the total octahedral stress at a depth of 1 ft (0.325 m). For a soil with a uniform unit weight, in-situ earth pressure coefficient = 1 and no

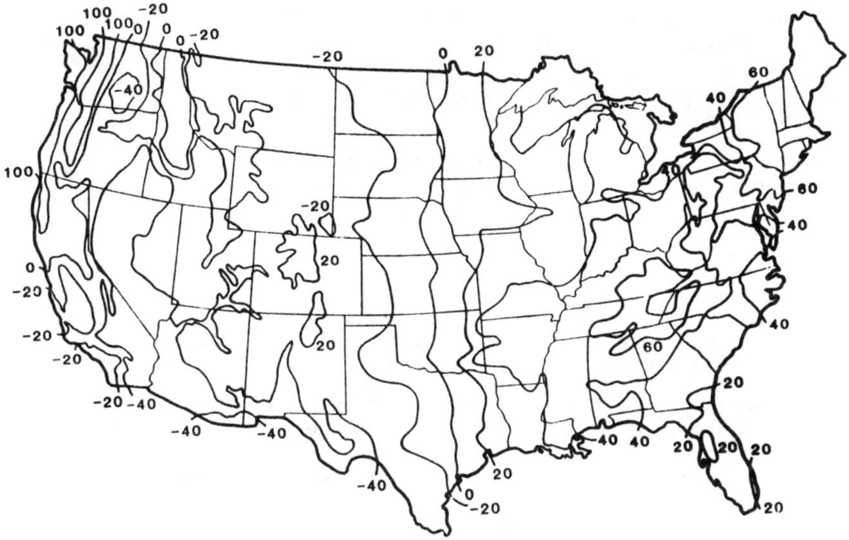

Fig. 5. Approximate Contours of Thornthwaite Moisture Index Across the United States (3)

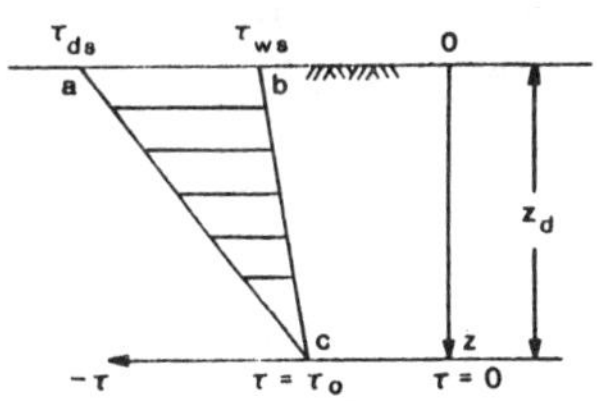

Fig. 6. Soil Suction Variation from Dry Season to Wet Season

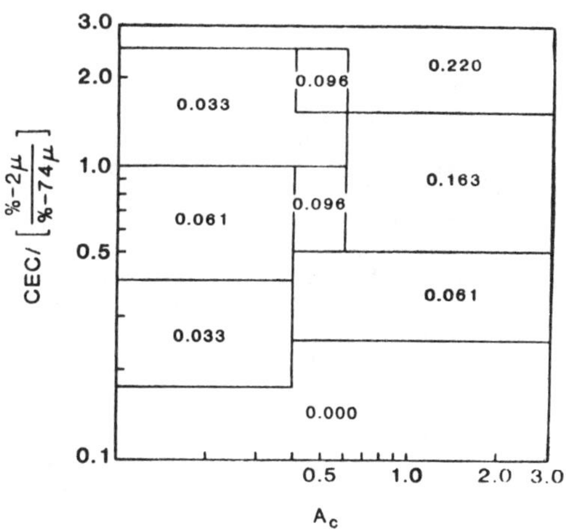

Fig. 7. Relationship for ξ_{100} (4)

external surcharge, σ_s/σ_1 = depth in feet. The following restrictions apply to the application of C (σ):

$$C(\sigma) = 0, \qquad \text{if } \log_{10}\left(\frac{\sigma_s}{\sigma_1}\right) \leq 0,$$

$$\text{and if} \quad C(\sigma) \geq \xi_\tau \log_{10}\frac{\tau_{final}}{\tau_{initial}}, \quad \left(\frac{\Delta H}{H_i}\right) = 0. \tag{9}$$

The term ξ_σ can be evaluated from

$$\xi_\sigma = \frac{C_r}{1 + e_o}, \tag{10}$$

where C_r = reload compression index ($\approx$ 0.15 C_c (normally consolidated soil)) and e_o = initial void ratio.

According to McKeen (4), ξ_τ can be obtained from the cation exchange capacity (CEC) and the activity of (A_c) the clay, where

$$CEC \approx [w_p(\%)]^{1.17} \qquad \text{(milliequivalents/100 gm)} \tag{11}$$

according to Mojekwu (5), and

$$A_c = \frac{I_p(\%)}{\left(\frac{\% - 2\mu}{\% - 74\mu}\right)(\%)}. \tag{12}$$

ξ_{100} is obtained from Fig. 7, and

$$\xi_\tau = \left(\frac{\% - 2\mu}{\% - 74\mu}\right) \times \xi_{100} \tag{13}$$

ANCHORAGE

Since the uplift movement requirement is established to prevent distress in the superstructure, it can be assumed that the relevant Q_T for design against soil expansion (Fig. 8) consists of approximately that part of the dead load that is applied to the shaft before structural continuity is established during construction of the superstructure = k DL, where DL = dead load. For purposes of this paper k is taken as 0.25; however, the load could be less than that value if a continuous system of grade beams is constructed before further construction of the superstructure. The soil heave profile and p', the thickness of the zone of partial uplift, is established using procedures described previously. The uplift capacity of the bell (if used) is established by assuming that the top of the zone of partial uplift is a free surface. A factor of safety of 2 is provided against uplift failure of the bell. Reliable suction at the base of the bell is unlikely to exist for significant periods of time and is therefore excluded.

The available downward-directed shear stress f_d in the stable zone for a straight-sided shaft is calculated from normal uplift capacity prediction procedures (e.g., 10). No allowance for reduced anchorage shear is made in just below the zone of partial uplift

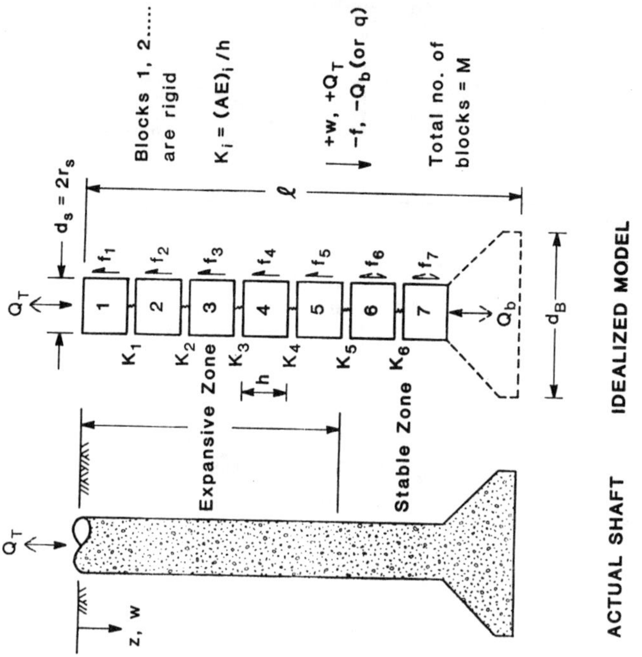

Fig. 9. Discretized Drilled Shaft

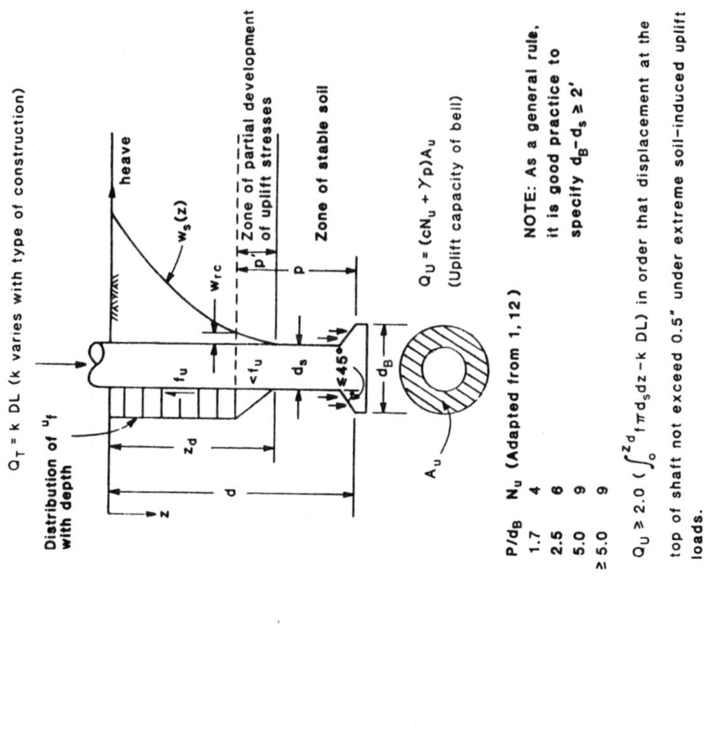

Fig. 8. Forces Acting on Drilled Shaft with Bell Anchor

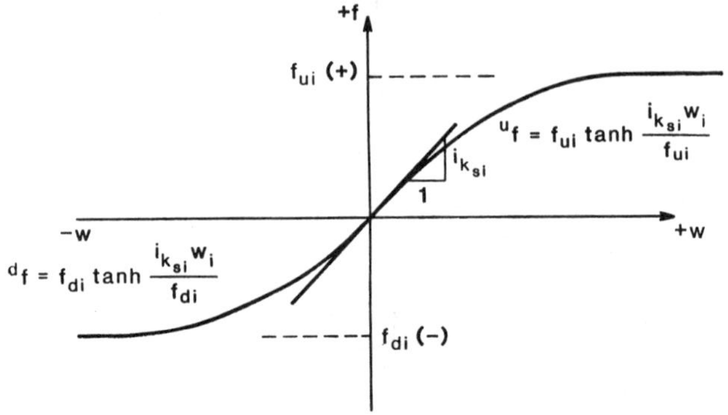

Fig. 10. Backbone f-w Curve: Node i

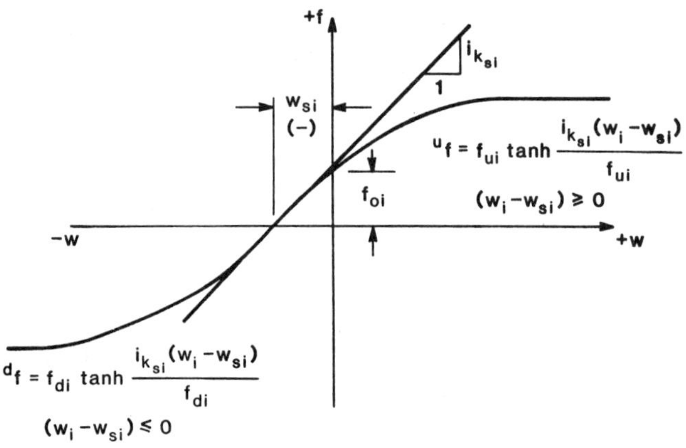

NOTE: q_c vs. w_M curve may be handled in a similar manner.

Fig. 11. Offset f-w Curve: Node i (Curve in Fig. 10 Translated by w_{si})

because this effect is accounted for in the normal uplift capacity procedure.

Since soil uplift loads can develop before any dead load is placed on the top of the shaft, tensile steel should be sized assuming $Q_T = 0$. Upward-directed Q_T, as might occur if soil uplift forces are allowed to act on grade beams or slabs before sufficient compression is applied to the shaft, must be minimized.

The critical section for reinforcement occurs at $z = z_d$ if the shaft is assumed rigid. If the shaft must be designed for uplift (negative Q_T), then the full design value of Q_T should be used in the static balance shown in Fig. 8.

BASIC CONSIDERATIONS FOR SOIL-STRUCTURE INTERACTION ANALYSIS

The simplified procedure described in the preceeding section is often adequate for the analysis of relatively short, rigid drilled shafts in expansive clays. However, for long flexible shafts or for shafts in which z_d is large, the zone in which the full development of f_u occurs is modified by the shaft movement, which is affected both by soil and external loading. Analysis of such a condition requires the use of a soil-structure interaction model. A relatively simple, approximate, practical model is described here that is based on the f-w curve concept. It can be used to investigate the relative importance of design parameters, which can be extremely beneficial in the design process.

Any practical procedure for analyzing drilled shaft load-deformation behavior in expansive clay should consider shaft flexibility, shear stresses due to uplift, anchorage or bearing behavior in the base (i.e., possibility of a bell), structural load, and nonlinear behavior of the soil. These features are depicted schematically in Fig. 8. A soil expansion profile is shown on the right side of the shaft, and the stresses on the shaft at the shaft-soil interface are indicated on the left side of the shaft. The relation $w_s(z)$ is the absolute free-field heave profile developed from suction changes. The method described here for defining the free-field heave profile from suction changes and depth to stable suction presumes that the soil surrounding the shaft remains under the influence of seasonal moisture variation. A downward displacement (w_s: free-field soil; or w: shaft) is considered positive throughout. f is a shear stress that is directed upward on the shaft (${}^uf$) above z_d (depth of seasonal moisture change) and directed downward (${}^df$) below z_d (not shown) if there is no applied load Q_T. When ${}^uf$ and ${}^df$ reach their absolute maximum values, they are denoted f_u and f_d. That is, the presuperscript denotes direction on the shaft, while the subscript denotes both direction and limiting value. q_{bu} is a bearing stress on the top of a bell, if present, if the integral of f on the perimeter of the shaft exceeds Q_T. Otherwise, a bearing stress q_{bc} (or no stress) exists on the base. The magnitudes of the stresses f and q depend on the magnitude of soil uplift, $w_s(z)$; z_d; shaft length; shaft flexibility; the presence of a bell; and the magnitude of Q_T. These factors also affect the deformation of the shaft.

ANALYTICAL MODEL

Fig. 9 presents a schematic of a one-dimensional discrete element model for computing the axial load-deformation behavior of a drilled shaft in expansive clay that incorporates the features just described. The cylindrical portion of the shaft is divided into M rigid blocks separated by springs, which simulate shaft stiffness by lumping the stiffness of each block at the interface between two blocks. The bell, if one is present, is considered to be rigid and is replaced by the upward or downward total soil reaction on the bell, Q_b. The blocks themselves are considered to be weightless. Each block may have a soil reaction acting on it in either direction. The sign convention and notation are shown in Fig. 9. Seven blocks are shown in the example; however, 15 to 20 are preferred in actual computations.

Considering static equilibrium of each block, the following stiffness equation can be written for the drilled shaft in matrix form:

$$\begin{vmatrix} -K_1 & K_1 & & & & & \\ K_1 & -(K_1+K_2) & K_2 & & 0 & & \\ & K_2 & -(K_2+K_3) & K_3 & & & \\ & & K_3 & -(K_3+K_4) & K_4 & & \\ & & & K_4 & -(K_4+K_5) & K_5 & \\ & 0 & & & K_5 & -(K_5+K_6) & K_6 \\ & & & & & K_6 & -K_6 \end{vmatrix} \begin{vmatrix} w_1 \\ w_2 \\ w_3 \\ w_4 \\ w_5 \\ w_6 \\ w_7 \end{vmatrix} = \begin{vmatrix} \pi f_1 h d_s - Q_T \\ \pi f_2 h d_s \\ \pi f_3 h d_s \\ \pi f_4 h d_s \\ \pi f_5 h d_s \\ \pi f_6 h d_s \\ \pi f_7 h d_s + Q_b \end{vmatrix} \quad . \; . \; . \; . \; . \; (14)$$

This equation cannot be solved directly since f_i and Q_b are not known. They are functions of the factors described in the preceding section. It is now assumed, therefore, that the soil at the level of each block possesses a two-branch hyperbolic relation between f and w, as shown in Fig. 10. The f-w relation for Block i is considered to be uncoupled from similar relations at all other blocks. A similar relation can be developed between the base bearing stress, q_b, and displacement of the bottom block, w_M.

The following inputs are needed for the hyperbolic unit load transfer curves. These inputs can vary from block to block, at the discretion of the analyst, to reflect variations perceived to exist physically. Analyses often can incorporate limiting or "bounding" conditions for purposes of arriving at a safe design.

${}^{i}k_{si}$ = initial soil subgrade modulus in shear at block i, F/L^3,

f_{ui} = absolute maximum shear stress in upward direction on shaft at block i, F/L^2,

f_{di} = absolute maximum shear stress in downward direction on shaft at block i, F/L^2,

^ik_b = initial subgrade modulus in bearing on the shaft base or bell, F/L^3,

q_u = absolute maximum bearing stress on bell in uplift, F/L^2,

q_c = absolute maximum bearing stress on base of shaft in compression, F/L^2.

q_c can be obtained from appropriate bearing capacity theory (e.g., 10), and q_u can be estimated from $q_u = cN_u + \gamma p$, where N_u can be obtained from Fig. 8; c is the average undrained shear strength from the base of the bell to a distance equal to $2d_b$ above the base of the bell; γ is the total unit weight of the soil and p (defined in Fig. 8) is estimated or computed from the free-field heave profile and w_{rc}. N_u should be reduced if the procedure used to determine shear strength fails to consider any form of premature failure that may develop, such as local failure or failure along joints in the soil.

The value of $^ik_{si}$ (initial soil subgrade modulus in shear for block i) can be obtained from Eq. 3 by observing that $^ik_{si} = f_{ui}/w_{rci}$. Therefore,

$$^ik_{si} = G_{si}/\{r_s[0.67 + \ln \frac{\zeta \ell}{r_s}]\}. \quad (15)$$

Factor ^ik_b can be evaluated empirically or from theory of elasticity. Using the logic first advanced by Skempton (9), it can be shown that, for clay soils in bearing,

$$^ik_b = \frac{N_c E_s}{d_b}, \quad (16)$$

where E_s is the initial Young's modulus of the clay in the region of the base of the shaft (evaluated with appropriate account given to the effects of stress relief during construction). E_s is defined, for example, as the secant modulus to a UU triaxial compression stress-strain curve at a ratio of stress difference to peak stress difference of about 5. Poisson's ratio of the soil is assumed to be 0.5. Eq. 16 can be applied for uplift (soil reaction on top of the bell) as well as bearing if $p \geq 2d_b$. For shallower bells ^ik_b must be reduced. Note that this method excludes soil suction. Therefore, if there is no bell, ^ik_b should be zero if w_M is negative (base moves upward).

When construction conditions are less than ideal (e.g., borehole left open for excessive periods and/or low-slump concrete used), these factors can assume values that are quite different from those that would be calculated from the procedures described above. Therefore, these procedures must not be considered as design rules but merely as guidelines that can be modified by the analyst based on experience, data from field tests and intuition concerning the soil mechanics aspects of the problem.

In problems involving uplift on bells, there is an additional element of shaft-soil interaction that is not considered explicitly in the method of analysis described here. The soil just above the bell will be pushed upward and outward by the bell, if the bell is drawn upward. This action will result in the development of lower shearing resistance between the shaft and soil just above the bell than would occur at the same relative shaft-soil movement if the bell were not present. In the analysis, therefore, it would be prudent to reduce f_d (or eliminate it altogether) for a distance equal to about $2d_b$ above the base of the bell.

A similar (but opposite) effect occurs in compression loading, especially for long-term loading. Downward movement of the bell removes support from the soil surrounding the shaft just above the bell, which allows the soil to move downward more easily under the shearing stress at the shaft-soil interface, thereby relieving that stress. For downward bell movement, therefore, it is also prudent to reduce f_u (or eliminate it altogether) for a distance equal to about d_b above the base of the bell.

No guidelines are given for bells in granular soils, as they would not normally be placed in such materials. For a straight-sided shaft that is terminated in sand, the generic base modulus developed from load tests (8) is

$$^{i}k_b = q_c/0.05\ d_b \tag{17}$$

where q_c(psi) = 8.3 N, and N is the uncorrected SPT blow count in blows per foot. N should not exceed 75.

The behavior of real soil is nonlinear, so that fixed moduli are in general insufficient to model shaft-soil interaction. Furthermore, there will be upward free-field movement of the soil relative to the shaft in the zone of expansion. If the unit load transfer curves (shaft or base) are assumed to shift along the shaft axis by an amount equal to the free-field soil uplift (or settlement) at node i (= w_{si}; negative for swelling soil) relative to their initial position, Fig. 11, a simple reformulation is possible:

$$f_i = f_{oi} + {}^{i}k_{si}\ w_i, \quad \text{and} \tag{18}$$

$$Q = Q_o + {}^{i}k_b\ A_b\ w_M, \tag{19}$$

where f_{oi} is the induced shear stress on the shaft, assuming temporarily that the shaft remains rigid but that free-field soil movement occurs; Q_o is the induced base load under similar conditions; $^{i}k_{si}$ is the initial tangent modulus to the f-w curve; $^{i}k_b$ is the initial tangent modulus to the q-w_M curve; and A_b is the area of the base.

If Eq. 14 is rewritten with these definitions of f and Q, the problem can be solved as long as the soil and shaft behavior remains linear (small displacements), since f_{oi} and Q_o can be evaluated from the expression for u_f on Fig. 11 (or corresponding expression for Q) with $w_i = 0$. "Solve the problem" means to obtain a solution for w_i for i = 1 to M for a given value of Q_T (positive, negative or zero). From this vector of displacements, forces in the shaft can be computed at each "spring" location (Fig. 9) from

$$Q_i = (w_i - w_{i+1})\ (AE)_i/h = [B(f_i+f_{i+1})/2], \qquad (20)$$

$$B = \pi h d_s \qquad (21)$$

If Q_i is of such magnitude as to produce concrete tensile failure, $(AE)_i$ should be modified appropriately and the analysis repeated.

For nonlinear problems (large displacements, applying to most expansive clay problems) an incremental-iterative approach is used. In this approach the external applied loads (and the induced loads from the offset unit load transfer curves) are divided into N equal increments (N = 10± for most problems). The problem is then solved N times using an incremental stiffness matrix equation (Eq. 22), which is derived from Eq. 14. The change in soil reaction due to shaft movement is represented by the modulus ${}^nk_{si}$ and ${}^nk_b$ for the nth increment. These moduli are updated for each increment by taking the derivative of the hyperbolic tangent function at the current value of w_i. These moduli are summarized in Eqs. 23-28.

$${}^nk_{si} = {}^ik_{si}\ \mathrm{sech}^2 \left| \frac{{}^ik_{si}({}^{n-1}w_i - w_{si})}{f_{ui}} \right| \qquad ({}^{n-1}w_i - w_{si}) > 0; \quad (23)$$

$${}^nk_{si} = {}^ik_{si}\ \mathrm{sech}^2 \left| \frac{{}^ik_{si}({}^{n-1}w_i - w_{si})}{f_{di}} \right| \qquad ({}^{n-1}w_i - w_{si}) \leq 0; \quad (24)$$

$${}^nk_b = {}^ik_b\ \mathrm{sech}^2 \left| \frac{{}^ik_b({}^{n-1}w_M - w_{sM})}{q_c} \right| \qquad ({}^{n-1}w_M - w_{sM}) > 0; \quad (25)$$

$${}^nk_b = {}^ik_b\ \mathrm{sech}^2 \left| \frac{{}^ik_b({}^{n-1}w_M - w_{sM})}{q_u} \right| \qquad ({}^{n-1}w_M - w_{sM}) \leq 0; \quad (26)$$

$$f_{oi} = f_{ui} \tanh \frac{{}^ik_{si}|w_{si}|}{f_{ui}}; \qquad (27)$$

$$q_o = q_c \tanh \frac{{}^ik_b|w_{sM}|}{q_c} \quad \text{(normally zero, if base is in stable soil);} \qquad (28)$$

where n = increment number.

$$\begin{bmatrix} -(K_1 + {}^nk_{s1}B) & K_1 & & & & & 0 \\ K_1 & -(K_1 + K_2 + {}^nk_{s2}B) & K_2 & & & & \\ & K_2 & -(K_2 + K_3 + {}^nk_{s3}B) & K_3 & & & \\ & & K_3 & -(K_3 + K_4 + {}^nk_{s4}B) & K_4 & & \\ & & & K_4 & -(K_4 + K_5 + {}^nk_{s5}B) & K_5 & \\ & 0 & & & K_5 & -(K_5 + K_6 + {}^nk_{s6}B) & K_6 \\ & & & & & K_6 & -(K_6 + {}^nk_{s7}B + {}^nk_bA_b) \end{bmatrix} \begin{bmatrix} {}^n\Delta w_1 \\ {}^n\Delta w_2 \\ {}^n\Delta w_3 \\ {}^n\Delta w_4 \\ {}^n\Delta w_5 \\ {}^n\Delta w_6 \\ {}^n\Delta w_7 \end{bmatrix} = \begin{bmatrix} -{}^n\Delta(Q_T - f_{01}B) \\ {}^n\Delta f_{02}B \\ {}^n\Delta f_{03}B \\ {}^n\Delta f_{04}B \\ {}^n\Delta f_{05}B \\ {}^n\Delta f_{06}B \\ {}^n\Delta(f_{07}B + q_bA_b) \end{bmatrix} \tag{22}$$

$${}^nw_i = \sum_{j=1}^{n} {}^j\Delta w_i \qquad B = \pi h d_s$$

${}^n\Delta(x)$ = Initial Induced External Load $(x) \div$ No. of Increments

A_b = Base Bearing Area (on Bottom)

Once the solution is achieved, the w-vector will no longer be null, and the assumption that initial relative displacement between soil and shaft equals w_s is no longer correct. The offset, w_{si} in Fig. 11, must be modified to $w_{si} - w_i$ and the analysis repeated. This process is followed iteratively until the w-vectors on two successive iterations are identical within a specified tolerance. A flow chart for the procedure described here is given in Fig. 12.

EXAMPLE

To illustrate the method, the solution to an example problem, described in Fig. 13, is given. From the soil and site information, a free-field heave profile is computed using Eqs. 4a-13, the ${}^ik_s$ values are computed from Eq. 15, and k_b is computed from Eq. 16, assuming $N_c = 9$ and $E_s = 3G_s$ (at base level). Average cracking, f = 0.8, is also assumed.

The results of the free-field swell calculations are given in Table 1, and the ${}^ik_s$ and k_b values are given in Table 2. The solution, in the form of steel and concrete stresses (compression positive) and displacement (downward positive) verses depth, is given in Fig. 14. The concrete stresses in this example design are of such a magnitude that tensile cracking is unlikely. The movement of the shaft is upward despite the application of the external compression load.

Table 1. Free-Field Swell Profile; Example Problem

Depth Interval (ft)	$\tau_{initial}$ (cm)	τ_{final} (cm)	$\frac{\sigma_s}{\sigma_1}$	$-\xi_\tau \log_{10} \frac{\tau_f}{\tau_i}$	$c(\sigma)$	ΔH (in.)	w_s (in.)
0-2	- 225000	- 2560	1	0.080	0	1.54	3.89
2-4	- 161500	- 2500	3	0.074	0.015	1.13	2.35
4-6	- 97900	- 2440	5	0.066	0.022	0.84	1.22
6-8	- 34200	- 2380	7	0.047	0.027	0.38	0.38

Table 2. Initial Interaction Moduli; Example Problem

Modulus	Value (pci)
${}^ik_s$ (all nodes)	58
k_b	450

Note: 1 ft = 0.305 m; 1 in. = 25.4 mm; 1 pci = 271.3 kN/m^3

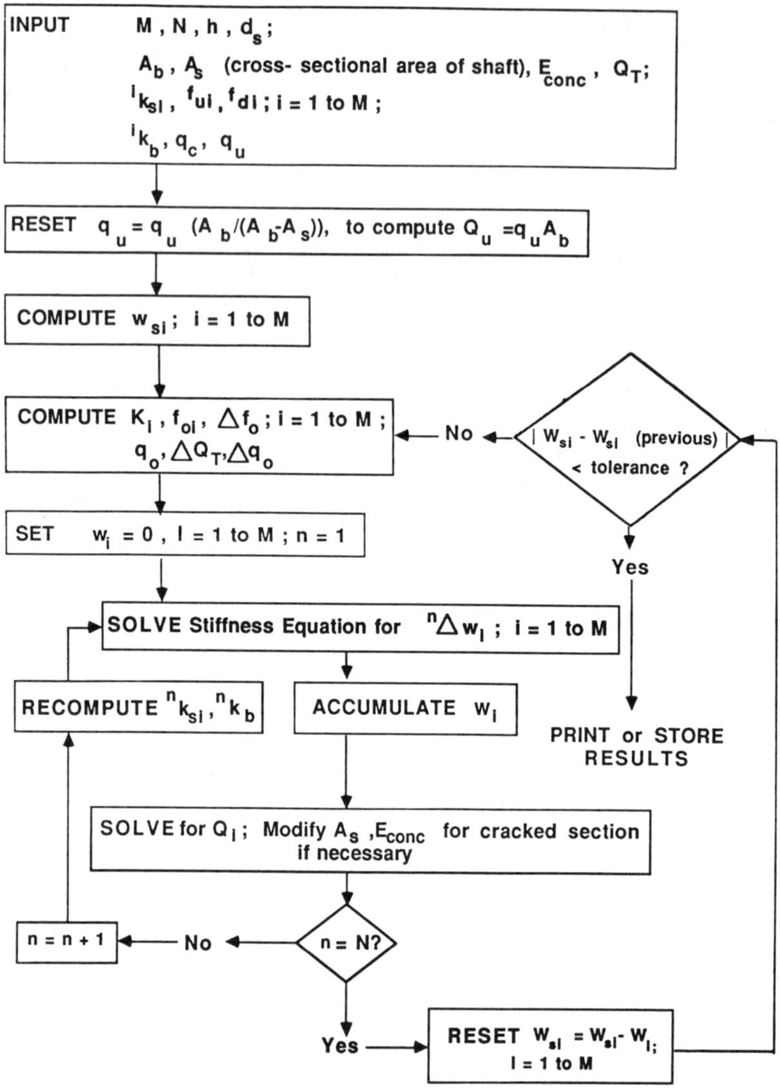

Fig. 12. Flow Chart for Drilled Shaft in Expansive Soil

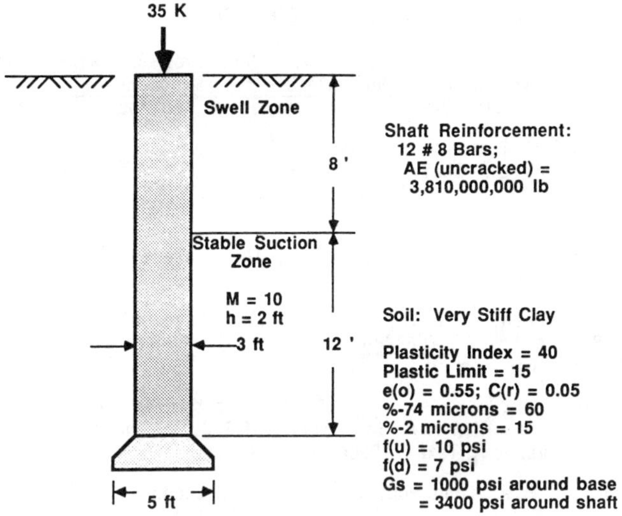

Fig. 13. Data for Example Problem

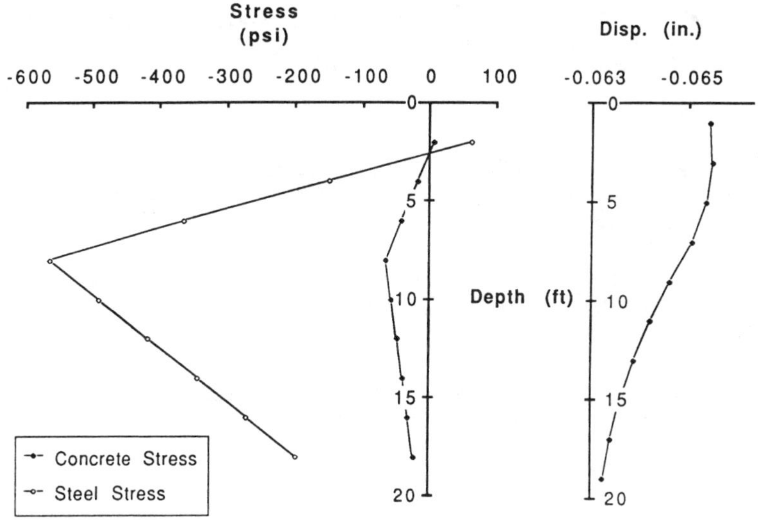

Fig. 14. Stress and Displacement vs. Depth; Example Problem

APPENDIX - REFERENCES

1. Ismael, N.F., and Klym, T.W., "Behavior of Rigid Piers in Layered Cohesive Soils," Journal of the Geotechnical Engineering Division, ASCE, Vol. 104, No. GT8, August, 1978, pp. 1061-1074.

2. Johnson, L.D., and Stroman, W.R., "Vertical Behavior of Two 16-Year-Old Drilled Shafts in Expansive Soil," Analysis and Design of Pile Foundations, J.R. Meyer, Ed., ASCE, October, 1984, pp. 154-173.

3. Lytton, R.L., PTI Design Manual, The Post-Tensioning Institute, Phoenix, AZ, 1979.

4. McKeen, R.G., "Design of Airport Pavements for Expansive Soils," Report No. DOT/FAA/RD-81/25, USDOT, Federal Aviation Administration, Washington, D.C., January, 1981.

5. Mojekwu, E.C., "A Simplified Method for Identifying the Predominant Clay Mineral in Soil," M.S.C.E. thesis, Texas Tech University, Lubbock, Texas, December, 1979.

6. O'Neill, M.W., and Poormoayed, N., "Methodology for Foundations on Expansive Clays," Journal of the Geotechnical Engineering Division, ASCE, Vol. 106, No. GT12, December, 1980, pp. 1345-1367.

7. Randolph, M.F., and Wroth, C.P., "Analysis of Deformation of Vertically Loaded Piles," Journal of the Geotechnical Engineering Division, ASCE, Vol. 104, No. GT12, December, 1978, pp. 1465-1488.

8. Reese, L.C., Touma, F.T., and O'Neill, M.W., "Behavior of Drilled Piers Under Axial Loading," Journal of the Geotechnical Engineering Division, ASCE, Vol. 102, No. GT5, May, 1976, pp. 493-510.

9. Skempton, A.W., "The Bearing Capacity of Clays," Proceedings, Building Research Congress, Division I, 1951, BRC, London, 1951.

10. Stas, C., and Kulhawy, F.H., "Critical Evaluation of Design Methods for Foundations Under Axial Uplift and Compression Loading," Report No. EPRI EL-3771, Electric Power Research Institute, Palo Alto, CA, November, 1984.

11. Thornthwaite, C.W., "An Approach Toward a Rational Classification of Climate," Geographical Review, Vol. 38, No. 1, 1948, pp. 55-94.

12. Yazdanbod, A., Sheikh, S.A., and O'Neill, M.W., "Uplift of Shallow Underreams in Jointed Clay," Foundations for Transmission Line Towers, J-L Briaud, Ed., ASCE, April, 1987, pp. 110-127.

Bearing Capacity and Settlement Characteristics of GEOWEB-Reinforced Earth Slabs

Vito A. Guido[1] M.ASCE

Sotirakis N. Christou[2]

ABSTRACT

Many soil reinforcing products are available to the geotechnical engineer today. One such product is GEOWEB-reinforcement, which is a three-dimensional grid (cell) confinement system. Grid confinement decreases lateral movement of the soil particles when loaded, thus increasing both stability and load carrying ability of low quality soils. Presented herein is a comparison of the results of laboratory model plate loading tests used to study the bearing capacity and settlement characteristics of GEOWEB-reinforced earth slabs. Four parameters were investigated: the number of layers of GEOWEB-reinforcement, the depth below the plate to the top of the first layer of GEOWEB-reinforcement, the size of the GEOWEB-reinforcement, and the relative density of the soil. For the tests performed; the bearing capacity of the earth slab increased, whereas, the settlement decreased with an increasing number of reinforcing layers, an increasing width of reinforcement, and a decreasing depth to the top of the first layer of reinforcement. In addition, the rate of increase of the bearing capacity and the rate of decrease of the settlement decreased with increasing relative density. At its maximum there was an 80% increase in bearing capacity and a 50% reduction in settlement due to the GEOWEB-reinforcement.

INTRODUCTION

An increased load bearing capacity and subsequent

-----

1. Associate Professor of Civil Engineering, The Cooper Union for the Advancement of Science and Art, School of Engineering, Cooper Square, New York, NY 10003.

2. Staff Engineer, KANIKA Constructions Ltd., P.O. Box 3029, Limassol, Cyprus.

settlement reduction of poor quality soils can be achieved through the use of reinforcement (geotextiles and polymer grids) which give the soil an added tensile strength. A bond between the soil and reinforcement is achieved either through friction (geotextiles) or interlock at the soil-reinforcemnt interface (polymer grids). Another type of reinforcement which utilizes the principle of lateral confinement is also currently available (GEOWEB). This paper will investigate the reinforcing capabililties of GEOWEB, and its effect on the bearing capacity and settlement characteristics of reinforced earth slabs. The results reported herein are a continuation of the ongoing research of the principal author on the effect of different reinforcing materials on reinforced earth slabs (Guido et al, 1985 and 1987).

GEOWEB GRID CONFINEMENT SYSTEM

The grid confinement concept was developed at the U.S. Army Engineer Waterways Experiment Station (WES), Vicksburg, MI. The concept of a three-dimensional soil confinement system was initially investigated using fabric lined gabions filled with sand (Webster and Watkins, 1977). Further tests were conducted using vertical lengths of corrugated plastic pipe, folding square aluminum grids fabricated from flat sheets, and resin impregnated hexagonal paper grids (Webster and Ashford, 1978 and Mitchell et al, 1979). The aluminum grids performed well but proved to be expensive and relatively heavy, while the paper cells deteriorated in moist conditions. Laboratory testing determined that an important parameter in the grid confinement system was the ratio of the cell width, B_1 to the cell depth, h. It was found that a value of approximately 1.0 was optimum. In addition, axisymmetric cell shapes worked best while plastic materials proved to be durable, inexpensive and easy to handle. The present GEOWEB grid confinement system is the result of this testing.

GEOWEBS are lightweight grids made from high density polyethylene strips having dimensions of 11 ft. (3.4m)x8 in. (23 cm)x0.05 in (1.3 mm), connected by ultrasonic welds at 13 in. (33 cm) spacings. Each unexpanded grid section is 5 in. (13 cm.) thick and expands to form a honeycomb arrangement of 561 cells covering an area of 8 ft. (2.5 m)x20 ft. (6m). See Figure 1 and Table I for typical dimensions and structural properties of the aforementioned GEOWEBS.

TABLE I

GEOWEB STRUCTURAL PROPERTIES

	English System	SI System
1. Expanded Dimension	8 ft.x20ft.x8in.	2.5m x 6m x 23cm
2. Collapsed Dimension	11ft. x 5in.x8in.	3.4m x 13cm x 23cm
3. Panel Thickness Nominal	0.047 in.	0.119cm
4. Weight	5.7 lb/yd^2	30.3N/m^2
5. Cell Area	41 $in.^2$	265 cm^2
6. Cell Seam Node Pitch	13 in.	33 cm
7. Welds/Seam	7	7
8. Seams Tensile Peel Strength	150 lbs.	667N
9. Installation Temperature Range	$-16^{o}F$ to $110^{o}F$	$-27^{o}C$ to $43^{o}C$

Polymer Material: High Density Polyethylene

Color: Black

Carbon Black Content: 2%

Chemical Resistance: Superior

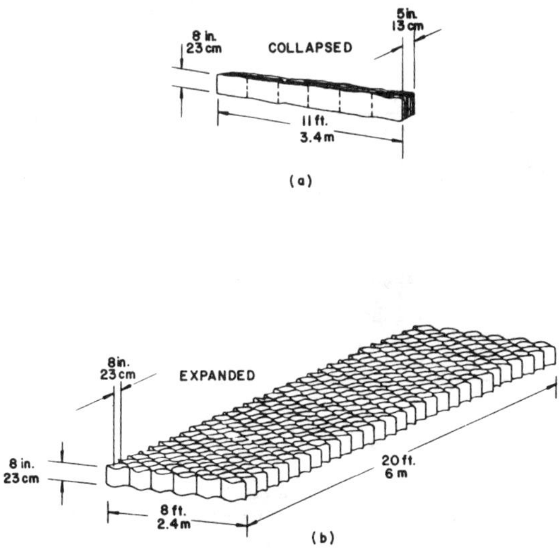

FIGURE 1 GEOWEB GRID CONFINENT SYSTEM
(a) COLLAPSED AND (b) EXPANDED

The GEOWEB grid confinement system not only increases the bearing capacity of the soil but also substantially reduces the settlement. This is accomplished by the confinenment of the failure wedges, which would be developed in an unreinforced soil, from displacing laterally and outward. In addition, the frictional interlock between the infill material and the cell walls allows the load to be distributed or shared with adjacent cells.

LABORATORY MODEL

The material presented herein is the result of 41 model plate loading tests. These tests were performed in a square stiffened wooden box, 48 in. (1.22m) wide and 36 in. (0.92m) deep. The soil used in all tests was a uniformly graded sand (SP). The properties of the sand can be found in Table II and Figure 2(a). Depending on the relative density the sand was placed in 1 in. (2.5 cm) to 3 in. (7.6 cm) lifts. In all but 9 tests the relative density was 55%. Since the GEOWEB grid confinement system works very well with poor quality soils, the use of a 55% relative density for the sand dramatically demonstrated the benefits to be derived from the GEOWEB. Subsequent tests, where the relative density was varied, illustrated that the benefits to be gained at higher relative densities were very small, see Figure 8.

TABLE II

PROPERTIES OF THE SAND USED IN MODEL TESTS

Uniformity Coefficient, Cu	1.90
Coefficient of Concavity, Cc	1.23
Effective Size, D_{10},mm	0.086
Specific Gravity	2.66
Minimum Dry Unit Weight, γ_{dmin}, kN/m^3	13.10
Maximum Dry Unit Weight, γ_{dmax}, kN/m^3	15.65
Dry Unit Weight, γ_d, kN/m^3	14.39

The GEOWEBS used did not have the dimensions indicated in Table I. The box used to perform the plate loading tests was too small to accommodate such large grids. A scaled-down version of the GEOWEB was used, retaining the cell width to cell depth ratio of 1.0 which had been found to be a critical parameter. Instead of the cell having a depth of 8 in. (20.3 cm) a depth of 2 in (5.1 cm) was used. In addition, the size of the collapsed GEOWEB was such that when expanded it would yield the desired size to be tested in multiples of the width of the loading plate. It was felt that the particle size of the sand used and the scaled-down grid size would be compatible producing minimal modeling effects on the rsults. The GEOWEB was stretched on a wooden frame, especially made for a specific width size of GEOWEB. This frame was subsequently removed when sand was placed inside the GEOWEB cells. Since the GEOWEBS used were 2 in. (5.1 cm) deep, enough sand was placed in the model, with the use of a mechanical drum dumper, to completely fill the cells and the annulus of space between the walls of the box and the GEOWEB-reinforcement. Depending on the desired relative density, the sand could have required vibration.

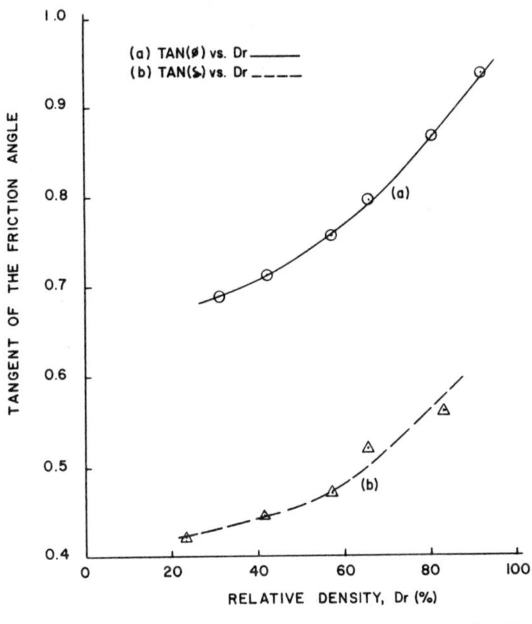

FIGURE 2 VARIATION OF TAN (ø) AND TAN (δ) WITH Dr.

The sand was loaded at its surface through a square plate with a 12 in. (30.5 cm.) width. The loading plate consisted of 0.75 in. (1.9 cm) exterior grade plywood stiffened by 2 - 0.3125 in. (8 mm) aluminum plates; one above and the other below the plywood. A static load was applied to the plate by a 25 ton (223 kN) capacity hydraulic cylinder and hand pump. The load applied in small increments, was recorded on a 20 ton (89kN) capacity load ring, and the resulting plate displacements recorded by two dial gages placed at the corners along a diagonal through the center of the plate. These dial gages had a 2 in. (5.1 cm) capacity with a 0.001 in. (25 μm) sensitivity. The GEOWEBS used were such that when expanded they were square in plan area and were placed concentrically under the square loading plate. No GEOWEB grid confinement system was used more than once to insure that no damaged reinforcemnt was used. The geometry of the test model is shown in Figure 3.

TEST RESULTS

The friction angle between the infill material and the cell wall, δ, is an important parameter since it allows the load to be distributed to adjacent cells and helps resist cell penetration. In order to determine this friction angle, a series of Fixed Shear Box Tests were performed.

FIXED SHEAR BOX TESTS

A series of 22 fixed shear box tests were performed using the sand in Table II and the high density polyethylene material of the GEOWEB confinement system. A square split shear box, 4 in. (10.2 cm) wide and 2 in. (5 cm) deep, was used. The GEOWEB material, attached to a wooden block, was placed in the lower half of the box and the sand was placed in the upper half of the box. The range of normal stresses used were from 13.9 psi (96 kPa) to 41.7 psi (287 kPa), while the relative densities ranged from 23% to 84%. A displacement rate of 0.0007 in./s (17.8 μm/s) was used. In Figure 2(b) the relationship between relative density and the tangent of the peak angle of friction between the sand and the GEOWEB material is shown. It can be seen that the friction angle developed between the sand and the GEOWEB material is substantially less than the friction angle developed within the soil itself. The efficiency, E,

$$E = \frac{\mathrm{TAN}\ (\delta)}{\mathrm{TAN}\ (\phi)} \qquad (1)$$

has an average value of approximately 0.644 for the range of relative densities considered. If the GEOWEB material were texturized in some manner the efficiency would be substantially increased. A testing program is currently investigating the effect of texturizing the GEOWEB material. Preliminary results indicate the efficiency can be increased to values of approximately 1.0.

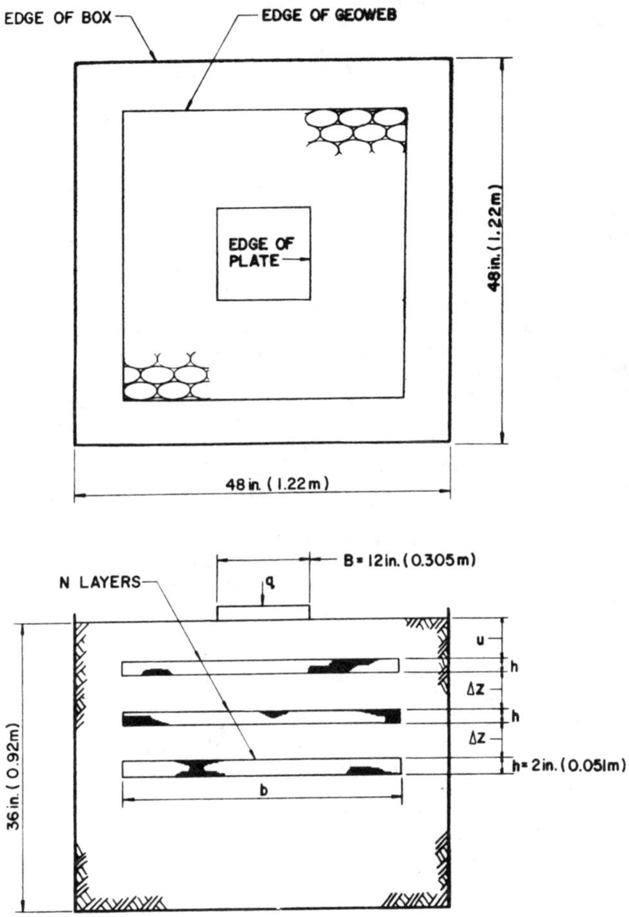

FIGURE 3 GEOMETRY OF TEST MODEL.

PLATE LOADING TESTS

The effect of four parameters on the bearing capacity and settlement characteristics of GEOWEB-reinforced earth slabs were investigated: They were:

1. Number of layers of GEOWEB-reinforcement, N.
2. Depth below the plate to the top of the first layer of GEOWEB-reinforcement, u, expressed in dimensionless form as u/B, where B is the width of the plate.
3. Width size, b, of a square area of expanded GEOWEB- reinforcement expressed in dimensionless form as b/B, the width ratio.
4. Relative density of the sand, D_r.

It should be noted that the vertical spacing, Δz, between the bottom of the previous layer of GEOWEB-reinforcement and the top of the next layer of GEOWEB-reinforcement was held constant in all tests at 3 in. (7.6 cm) or $\Delta z/B=0.25$ (See Figure 3). For the plate loading tests discussed herein, the depth of the GEOWEB cell used was 2 in. (5.1 cm) and the square plate used had a width of 12 in. (30.5 cm). In an actual design situation, the depth of the prototype GEOWEB would be 8 in. (23 cm) (see Table I) and the width of the footing would be considerably larger than 12 in. (30.5 cm). Therefore, $\Delta z/B$ values greater than 0.25 were considered impracticable, since large depths of embedment of the GEOWEB-reinforcement would result in an actual design situation.

For convenience in expressing and comparing test data, thebearing capacity ratio (BCR) and settlement reduction factor (SRF) will be used:

$$BCR = q_r/q_o \qquad (2)$$

and

$$SRF = (s/B)_r/(s/B)_o \qquad (3)$$

where q_o is the ultimate bearing pressure for the unreinforced sand at a specific settlement ratio, s/B and q_r is the bearing pressure of the GEOWEB-reinforced sand at a settlement ratio corresponding to the settlement ratio at the ultimate bearing pressure for the unreinforced sand. For the unreinforced case at a relative density of 55%, the load-settlement curve approached a vertical aymptote at a value of the settlement ratio, s/B, of 0.07. Failure was defined at this value of s/B, since an increase in settlement was not accompanied by as large a percent increase in load. $(s/B)_o$ is the settlement ratio for the unreinforced sand corresponding to a specific ultimate bearing pressure, q and $(s/B)_r$ is the settlement ratio for the GEOWEB-reinforced sand at a bearing pressure corresponding to the ultimate bearing pressure of the unreinforced sand. For a listing of values of q see Table III. Typical load-settlement curves are shown in Figure 4 for different values b/B.

TABLE III

VALUE OF BEARING PRESSURE

Dr%	q_o,psi (kPa) @ s/B =$(s/B)_o$=0.07	q_r,psi (kPa) @s/B=$(s/B)_o$=0.07
55	21.6 (148.9)	39.2 (270.3)
65	31.8 (219.2)	43.4 (299.2)
75	46.7 (322.0)	48.4 (333.7)
85	57.0 (393.0)	59.0 (406.8)

(*) u/B=0.25, Δz/B=0.25, N=3 and b/B=2.0.

1. Number of Layers of GEOWEB-Reinforcement

Figure 5 shows the variation of the BCR and SRF with the number of layers of GEOWEB-reinforcement. There is a substantial increase in the bearing capacity coupled with a substantial decrease in the settlement due to the addition of only one layer of GEOWEB-reinforcement. Subsequent increases in the bearing capacity and decreases in the settlement due to additional layers of GEOWEB-reinforcement result in diminishing values, approaching optimum values at approximately N=4 (z/B=1.67). This occurs well beyond the effective depth (z/B=1) observed by (Guido et al, 1985 and 1987) for geotextiles and geogrids. This is due to the three-dimensional nature of the GEOWEB-reinforcement, whereas, the geotextile and polymer grid types of reinforcement are two-dimensional.

2. Depth to the Top of the First Layer of Geoweb-Reinforcement

The variation of the BCR and SRF with u/B is shown in Figure 6. As the depth to the top of the first layer of GEOWEB-reinforcement decreases the BCR increases and the SRF decreases, indicating the bearing capacity increases while the settlement decreases. Similar results in BCR have been observed by (Akinmusuru and Akinbolade, 1981 and Guido et al, 1985 and 1987). At a value of u/B of approximately 1.0 the BCR is approaching a minimum value of 1.0 indicating the GEOWEB-reinforced earth slab is behaving as one which is unreinforced. In addition, the settlements associated with this u/B of 1.0 are essentially those of an unreinforced earth slab. A comparison of Figures 5 and 6 would yield that if the bottom of the

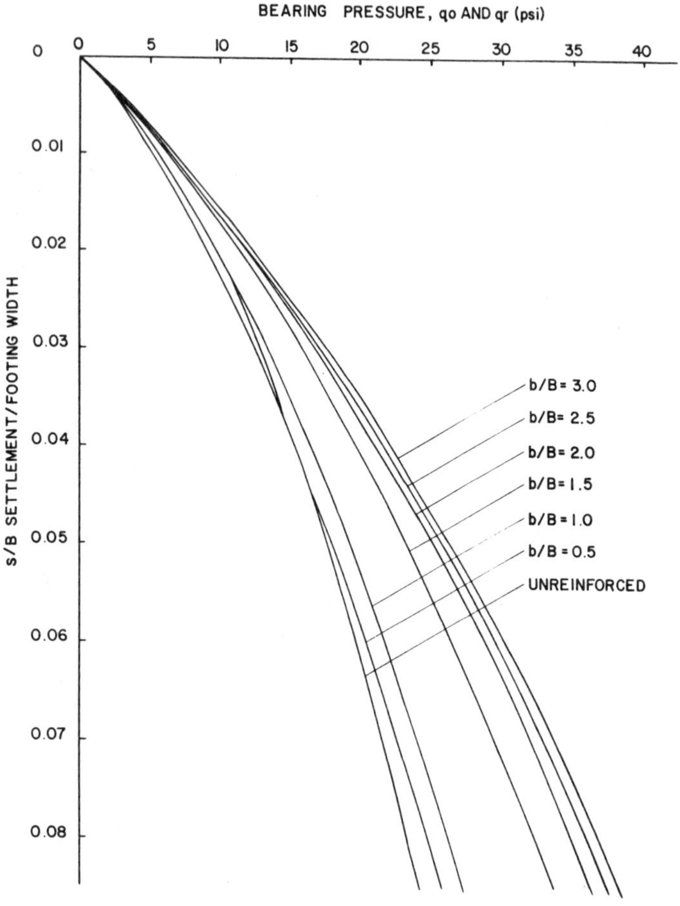

FIGURE 4 LOAD-SETTLEMENT CURVES FOR u/B=0.5, Δz/B=0.25, N=3 AND Dr=55%.

last layer of GEOWEB-reinforcement were at z/B=1.67 from Figure 5 (u/B=0.25 and N=4) the BCR and SRF are approximately 1.55 and 0.6, respectively, and from Figure 6 (u/B=1.0 and N=2) the BCR and SRF are approximately 1.05 and 1.0, respectively. This indicates that a high concentration of GEOWEB-reinforcement within the same depth of sand sufficiently stiffens the earth slab to produce substantially higher bearing capacities and lower settlements. Therefore, when u/B is sufficiently large the GEOWEB-reinforcement does not interfere with the formation of the failure planes and a shear failure of the sand occurs above the top of the uppermost layer of GEOWEB-reinforcement. The GEOWEB reinorcement acts as a rigid boundary preventing the shear zone to penetrate beyond it. This is one of the three failure modes of reinforced earth slabs described by (Binquet and Lee, 1975).

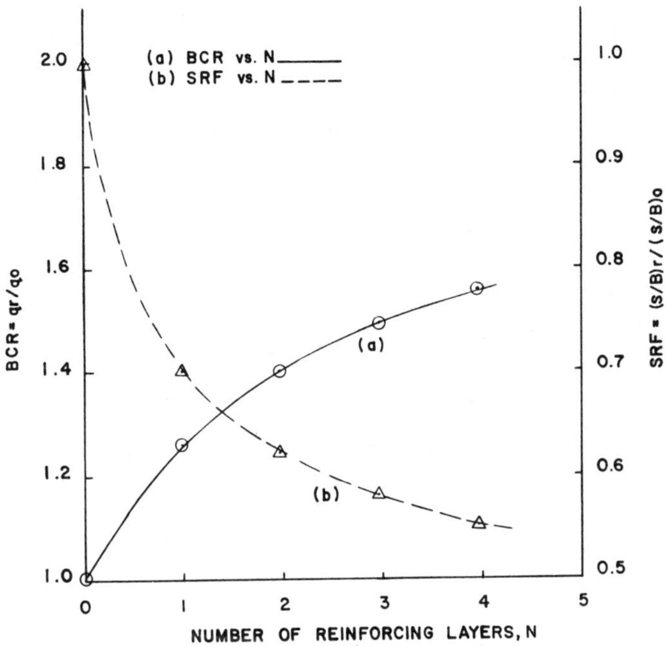

FIGURE 5 VARIATION OF BCR AND SRF WITH THE NUMBER OF LAYERS, U/B=0.25 ΔZ/B=0.25, b/B=2.0 AND Dr=55%.

3. Width Size of a Square Area of Expanded GEOWEB-Reinforcement

Figure 7 shows the variation of BCR and SRF width b/B. As b/B, increases the BCR increases and the SRF decreases, reaching a maximum value of BCR and a minimum value of SRF at a b/B value of approximately 2.0. Similar results have been observed by (Fragaszy and Lawton, 1984 and Guido et al, 1985 and 1987) for

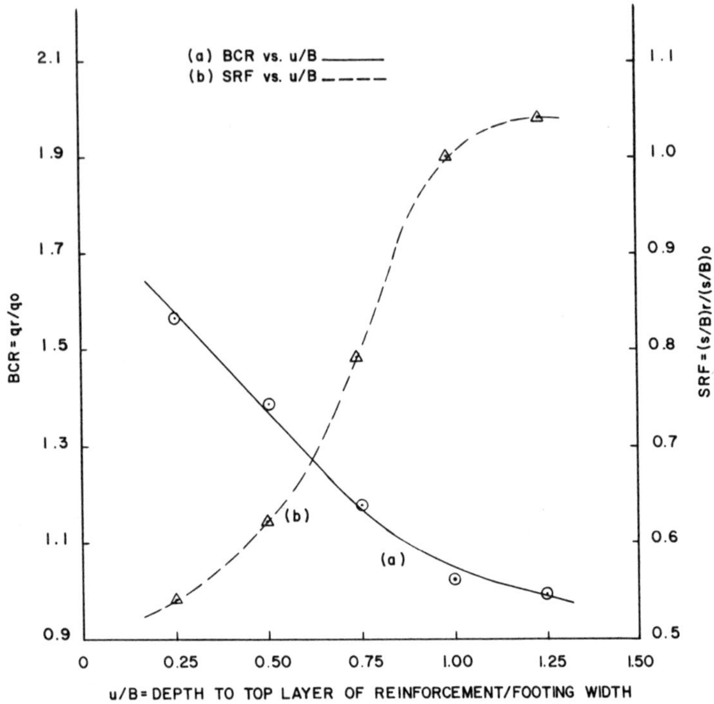

FIGURE 6 VARIATION OF BCR AND SRF WITH DEPTH TO TOP LAYER, $\Delta Z/B = 0.25$, $N = 2$, $b/B = 2.0$ AND $Dr = 55$ %.

different reinforcing materials. This indicates that to have an expanded GEOWEB of square area with a width larger than 2.0 times the width of the footing would be ineffective. The grid confinement system functions with the grid cells sustaining the imposed load. As the GEOWEB-reinforcement falls farther and farther outside the stress boundaries developed within the earth slab, the cells outside the stress boundaries sustain less and less of the load with the majority of the load being carried by the cells within the stress boundaries. For practical reasons, such as footing pit excavations, it may not be feasible to use a GEOWEB which expands to a square area with a width 2.0 times the width of the footing. Since the largest percent gain in bearing capacity and percent loss in settlement occur over a range of width ratios between 1.0 and 1.5, the dimensions of the footing pit should perhaps be based on this range of b/B values rather than 2.0.

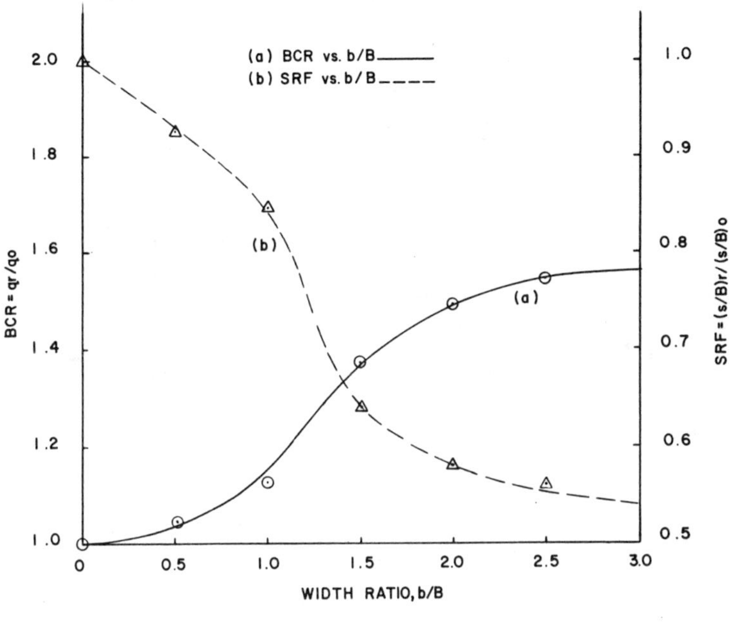

FIGURE 7 VARIATION OF BCR AND SRF WITH WIDTH RATIO, u/B=0.5, ΔZ/B=0.25, N=3 AND Dr=55%.

4. Relative Density of the Sand

As can be seen from Table III as the relative density increases from 55% to 85% there is a steady increase in bearing capacity from 21.6 psi (148.9 kPa) to 57.0 psi (393.0 kPa) for the unreinforced sand, and a steady increase in the bearing capacity from 39.2 psi (270.3 kPa) to 59 psi (406.8 kPa) for the GEOWEB-reinforced sand. It is evident that the percent gain in bearing capacity is substantially greater for the unreinforced sand than the GEOWEB-reinforced sand. The variation in BCR and SRF with relative density is shown in Figure 8. The GEOWEB-reinforced sand at a relative density of 55% shows an increase in bearing capacity of approximately 80% and a decrease in settlement of approximately 55% from that of the unreinforced sand, whereas, at a relative density of 85% the GEOWEB-reinforced sand shows an increase in bearing capacity of approximately 5% and a decrease in settlement of approximately 5% from that of the unreiforced sand. This indicates that for poorer quality soils tremendous benefits can be had from the

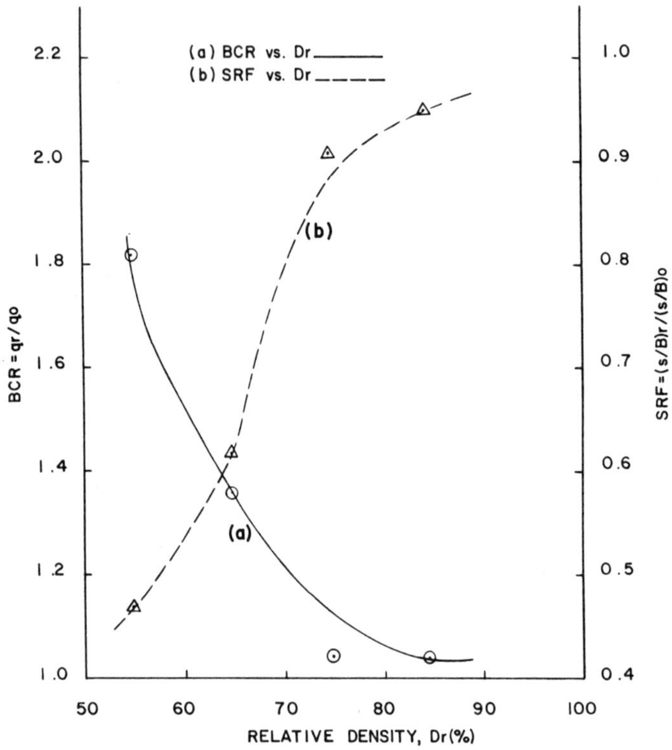

FIGURE 8 VARIATION OF BCR AND SRF WITH RELATIVE DENSITY, u/B=0.25, △Z/B=0.25, N=3 AND b/B=2.

inclusion of GEOWEB-reinforcement, whereas, for better quality soils the benefits may only be marginal. At lower relative densities the GEOWEB grid confinement system helped the earth slab to be more rigid than the unreinforced case, therefore, preventing lateral and outward displacement of the sand. This in turn leads to higher load bearing capacity and lower vertical displacemnts for the GEOWEB-reinforced earth slab versus the unreinforced earth slab. For higher relative densities the GEOWEB grid confinement system did not rigidify the earth slab substantially more than the already rigid unreinforced case, therefore, yielding essentially the same load bearing capacities and settlements for the reinforced and unreinforced cases.

CONCLUSIONS

This paper has discussed a series of laboratory model plate loading tests on GEOWEB-reinforced earth slabs. The results have shown that the insertion of GEOWEB-reinforcement into an earth slab can substantially increase its bearing capacity and decrease its settlement from those of an unreinforced earth slab. The GEOWEB-reinforcement should be placed within an effective depth (z/B=1) below the base of the footing and within the stess boundaries developed in the soil in order to obtain its maximum benefits. In addition, the relative density of the soil need not be high to achieve these maximum benefits. The soil can be placed in the cells at low relative density yielding large increases in bearing capacity and large decreases in settlement. One for one, the the presence of GEOWEB below the footing provides little improvement when compared with the situation where the same amount of well-compacted backfill is placed with no GEOWEB. The advantage of using GEOWEB below the footing occurs when an inferior grade backfill material is to be used (such as a poor quality in-situ soil) instead of quality backfill material, and comparable improvements are obtained at much reduced construction costs. It is the authors' belief that texturizing of the GEOWEB material within the cell will additionally improve the load bearing capacity and settlement characteristics of a GEOWEB-reinforced earth slab. Currently research is being conducted to investigate the effect of texturizing. Preliminary results indicate a substantial increase in the angle of friction between the GEOWEB material and the soil for the texturized material. Since friction between the infill material and the cell wall helps to transfer load from cell to cell this is an important development. Therefore, the amount of texturizing of the GEOWEB material is yet another important parameter to consider.

APPENDIX. - REFERENCES

1. Akinmusuru, J.O. and Akinbolade, J. A. 1981. "Stability of Loaded Footings on Reinforced Soil", J. Geotech. Engrg., ASCE, Vol. 107 (6), pp. 819-827.

2. Binquet, J. and Lee, K. L. 1975. "Bearing Capacity Analysis of Reinforced Earth Slabs", J. Geotech. Engrg., ASCE, Vol. 101 (12), pp. 1241-1255.
3. Fragaszy, R. J. and Lawton, E. 1984. "Bearing Capacity of Reinforced Sand Subgrades", J. Geotech. Engrg., ASCE, Vol. 110(10), pp. 1500-1587.
4. Guido, V.A., Bisiadecki, G. L., and Sullivan, M.J. 1985. "Bearing Capacity of a Geotextile-Reinforced Foundation". Proceedings of the 11th International Conference on Soil Mechanicsand Foundation Engineering, Vol. 3, San Francisco, CA pp. 1777-1780.
5. Guido, V. A., Knueppel, J. D., and Sweeney, M. A. 1987. "Plate Loading Tests on Geogrid-Reinforced Earth Slabs".Proceedings of the Geosynthetics Conference '87,Vol. 1, New Orleans, LA. pp.216-225.
6. Mitchell, J. K., Kao, T. C. and Kavazanjian, E. 1979. "Analysis of Grid Cell Reinforced Pavement Bases", U. S. Army Waterways Experiment Station, Technical Report GL-79-8.
7. Webster, S. L. and Ashford S. J. 1978. "Investigation of Construction Concepts for Pavements Across Soft Ground", U.S. Army Waterways Experiment Station, Technical Report S-78-6.
8. Webster, S. and Watkins, J. E. 1977. "Investigation of Construction Techniques for Tactical Bridge Approach Across Soft Ground", U.S. Army Engineer Waterways Experiment Station, Technical Report S-77-1.

Soil Modulus from Large Scale Plate Load Test

A. Holeyman[1)], M. Bottiau[2)]
and K.R. Massarsch[(3)]

ABSTRACT

An innovative plate load test has been developed, using the 44 tons (400 kN) load provided by a piling rig. Because of the large size of the loading plate (80 ft^2 or 7.5 m^2), the depth of influence typically extends to about 20 ft (6 m). The piling rig is instrumented with sensors to measure the load-deformation characteristics of the soil. Simple charts are produced for the interpretation of field data in terms of reaction modulus.

An extensive field testing programme was performed to check the relative soil improvement achieved by sand column installation. The comparison between moduli determined empirically from CPT and PMT tests and those measured by the large plate load test shows good agreement and that settlements can be significantly reduced by the installation of displacement-type sand columns.

INTRODUCTION

The efficiency of stone columns is affected by the method of installation. One common method uses an uncased hole which is created by water-jetting and the column is constructed by compacting the stone which is added from the ground surface, by a vibrator. Displacement type stone columns are installed in a similar way as conventional driven in-situ cast piles. A steel casing is driven into the soil, and then gradually withdrawn while the stone, gravel or sand is effectively compacted by a free-fall hammer. In this way the strength and the shape of the shaft can be controlled.

1) Deputy General Manager, 2) Project Engineer, 3) General Manager Technology; S.A. Franki, 196, Rue Grétry, 4020 Liège, BELGIUM

The compaction effect of stone column installation is normally checked by conventional geotechnical methods such as static cone penetration tests (CPT), Standard Penetration Tests (SPT) or pressuremeter tests (PMT). However, it is difficult to predict from tests performed at individual points the load-deformation behaviour of the soil mass which is reinforced by the stone columns.

In order to overcome this difficulty, a method has been developed, where the weight of the piling rig is utilized to carry out a large-scale plate load test, called a static ground test (SGT). SGT can be interpreted in two ways, either as a qualitative test, where only the increase in soil stiffness is measured, or as a quantitative test, where the actual modulus of subgrade reaction, k, is determined. As a static ground test is carried out within a short time period (several minutes), excess pore water pressure can not dissipate in cohesive soils and thus the soil will be tested under undrained loading conditions. However, as the stone columns consist of granular material, these will be loaded under drained conditions.

An extensive field testing programme was performed on a site in the vicinity of Antwerp, Belgium in order to assess the performance of displacement-type sand columns. The objective was to measure in the field the effect on the soil stiffness of sand columns and of a surface fill with varying thickness. This paper summarizes the test results which allowed us to compare the soil stiffness deduced from the literature, and from conventional geotechnical investigations, with that determined from large-scale plate load tests (SGT).

The successive stages of the field testing programme involved :

a) initial CPT and PMT,
b) static ground testing (SGT) of the virgin site, of the virgin site covered with a 16 in (40 cm) sand layer, of the same site after installation of 26 ft (8 m) deep sand columns and after placement of an additional 12 in (30 cm) lift of sand,
c) CPT, PMT in and between sand columns and small-scale plate load tests.

Fig. 1.a illustrates the grid pattern of the 19 sand columns and fig. 1.b the execution of the test programme.

GEOTECHNICAL CONDITIONS

The geotechnical conditions on the site are characterized by an about 2.6 ft (0.8 m) thick fill on an at least 42.3 ft (13 m) deep deposit of grey sandy clay. The water table is about 6.5 ft (2 m) deep.

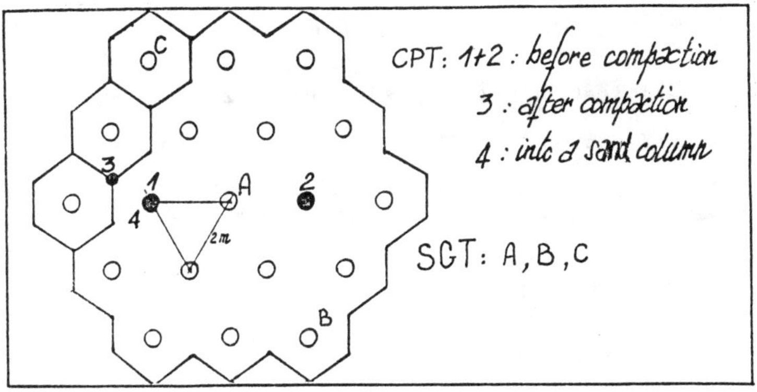

Fig. 1.a. Grid pattern of stone columns

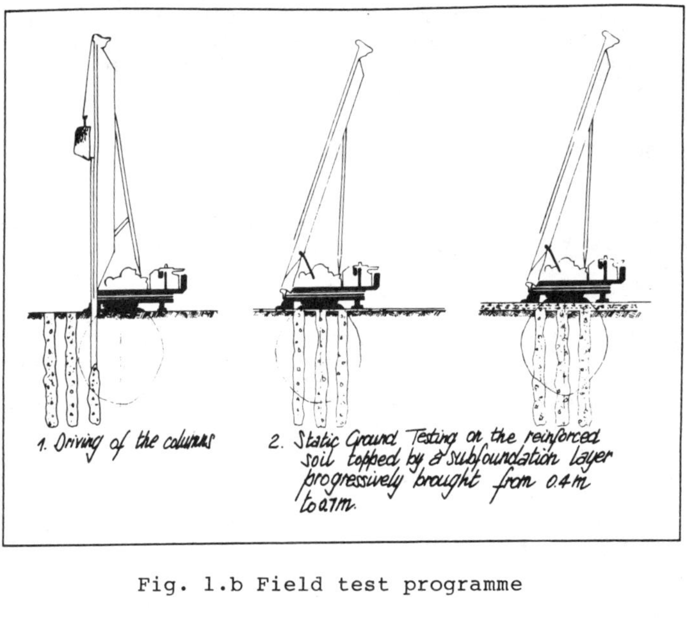

Fig. 1.b Field test programme

The geotechnical properties of this clay are well known in Belgium, belonging to the Rupelian formation (tertiary). The clay is stiff, overconsolidated and fissured with a water content close to the plastic limit. The cone resistance (electric cone) varies typically between 100 psi (1) and 400 psi (4 MPa) and the sleeve resistance yields an average friction ratio of 5.5 %.

Van Burm and Maertens (6) report the following typical properties :

Classification : CH PI = 52

2 μm fraction ≃ 49 % w_ℓ = 81 %, w_p = 29 %

The top layer (about 8 m) of this clay has the following strength and deformation properties :

Undrained shear strength :
Unconfined compression test : 75 kPa (10.9 psi)
Field vane test : 250 kPa (36.3 psi)
Elastic (Young's) Modulus :
E_{50}: Undrained conditions : 20 MPa (2003 psi)

In an overconsolidated clay, the installation of displacement-type sand columns will have little effect on the soil sitffness. Thus the soil improvement will be mainly caused by the reinforcing effect of the compacted stone columns. Previous investigations have shown that in soils with a friction ratio exceeding 3 %, no compaction effect can be achieved (2).

LARGE-SCALE PLATE LOAD TEST

The Franki walker-type rig moves by alternate movements of a large central plate and two adjacent smaller plates (beams), cf. Fig. 2. The weight of the rig (400 kN, 89 klb) is transferred by the 7,5 m^2 (78 ft^2) large central plate to the soil. Because of the comparatively large size of the plate, the depth of influence typically extends to about 6 m (19.38 ft). For the test programme, a conventional piling rig was instrumented to measure continuously the load-deformation characteristics of the soil. Displacement transducers were used to measure the relative displacement between the soil below the central plate and reference beams. Thus the relative displacements between the central plate and the adjacent beams is recorded. A pressure transducer measured the oil pressure applied to the jacks and thus a value proportional to the contact pressure at the central plate-soil interface, cf. Fig. 2. The test provides load-settlement curves for a stress range of 0 to 50 kPa (0-7.25 psi).

An example of such a curve is shown in Fig. 3. It should be noted that when the applied pressure reaches the maximum contact pressure, the reference beams are lifting off the ground, which results in large apparent deformations, cf. Fig. 3. These curves allow to determine the apparent secant modulus k_T for a contact stress range e.g. 20-40 kPa (2.9 - 5.8 psi), as illustrated in Fig. 4.

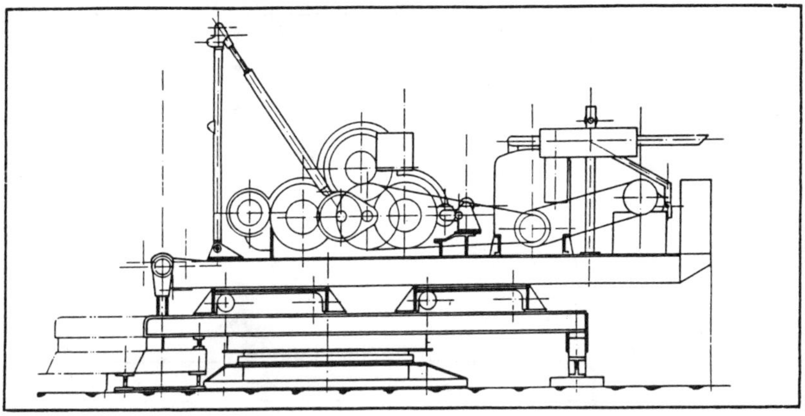

Fig. 2.a Sketch of walking mechanism of the piling rig.

Fig. 2b Franki rig instrumentation with data acquisition system

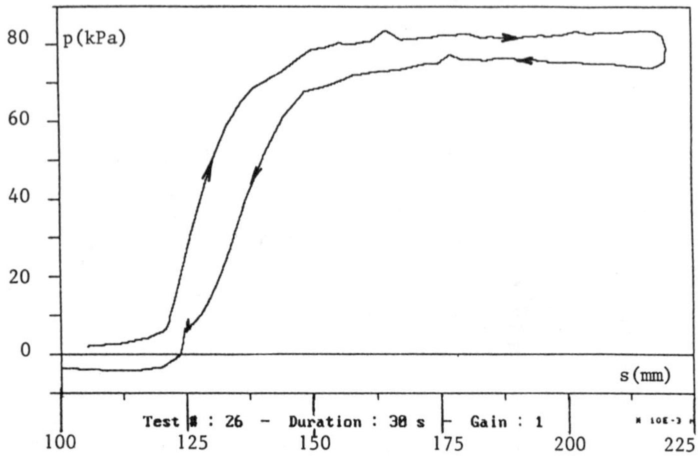

Fig. 3. Stress-displacement curve (1 kPa = 0.145 psi, 1 mm = 0.0394 in)

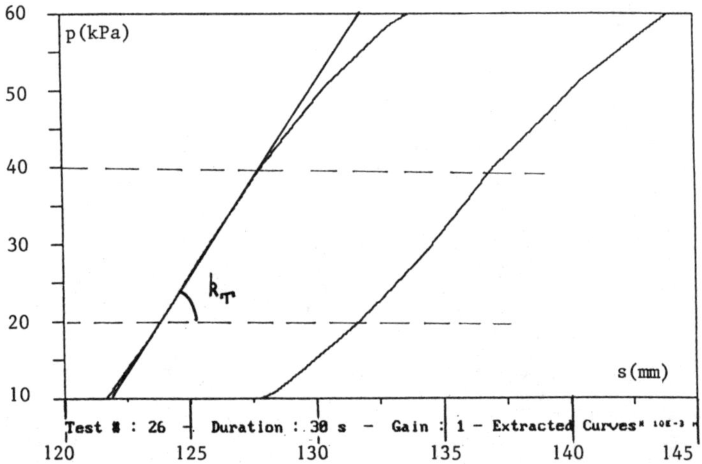

Fig. 4. Enlarged section of the load-deformation curve for detailed data interpretation (1 kPa = 0.145 psi, 1 mm = 0.0394 in)

where: $$k_T = \frac{\Delta p}{\Delta S} \qquad (1)$$

with Δp : increase of contact pressure
ΔS : increase of differential settlement between central plate and beams.

Also the flexibility of the rig chassis needs to be taken into account, either by structural analysis of the steel frame or, preferably by field calibration.

Test Interpretation

The absolute displacement of the central plate can either be measured (with respect to a reference point) or estimated using theory of elasticity. The relative movements between the central plate and the beams can then be determined by comparing the two loading conditions as illustrated in Fig. 5:

a) the weight of the rig is fully carried by the small beams without contact pressure below the central plate

b) the weight of the rig is fully carried by the central plate while the beams begin to lift off the ground

This analysis involves the calculation of settlements on an elastic half-space, resulting from the above loading conditions. These settlements have been evaluated assuming an elastic, homogeneous soil according to Poulos and Davis (4) for the geometries of the central plate and of the beams cf. Fig. 6.

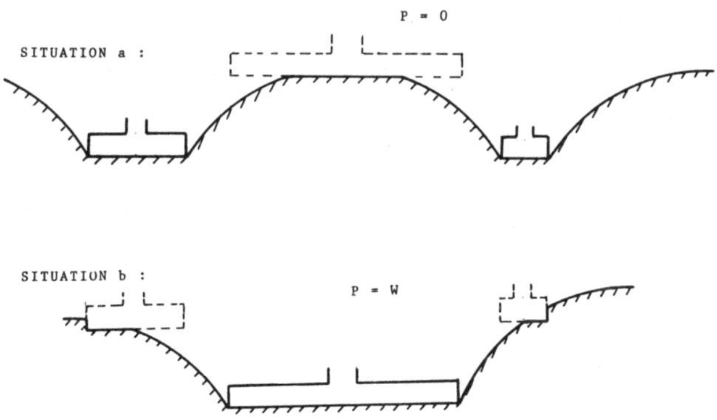

Fig. 5. Loading conditions below rig.

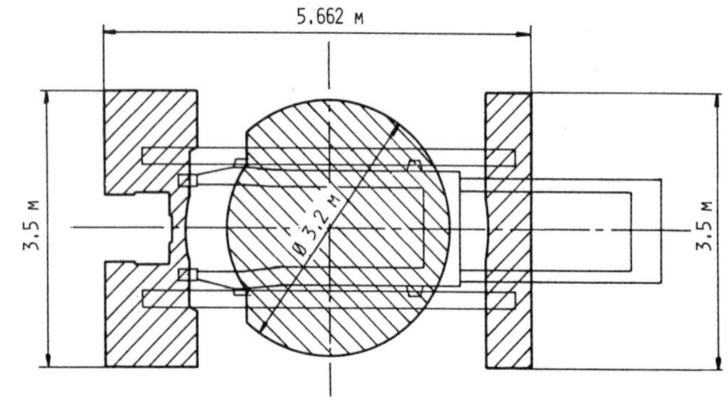

Fig. 6. Geometry of the central plate and outer beams.

The results are presented in a simple diagramme for deduction of the plane strain modulus E' :

$$E' = \frac{E_s}{1 - \nu_s^2} \tag{2}$$

with E_s : Young's modulus
ν_s : Poisson's ratio

Fig. 7 shows in full line the relation between E' and k_T assuming rigid plates. Alternatively, the coefficient of subgrade reaction k for the central plate can be determined from :

$$k + k_R = \beta\ k_T \tag{3}$$

with : β = 1.19 for the rigid plate of the Franki rig.
k_R = stiffness of the rig frame, as determined from field calibration.

For a circular rigid footing of radius R, the coefficient of subgrade reaction can be expressed by

$$k = \frac{2E'}{\pi R} \tag{4}$$

DETERMINATION OF COEFFICIENT OF SUBGRADE REACTION, K

Static load tests (SGT) were performed during the following stages of the test :

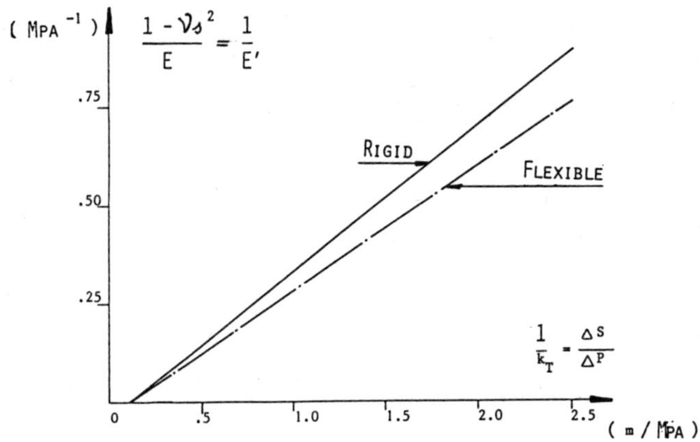

Fig. 7. Chart for direct interpretation of field tests. (1MPa = 145 psi, 1 m = 0.328 ft)

1 - on virgin soil
2 - on virgin soil provided with a 1.31 ft (0.4 m) thick fill.
3 - after installation of 26 ft (8 m) deep sand columns with a diameter of 1.62 ft (0.5 m) and a spacing of 6.46 ft (2 m).
4 - thickness of the surface layer of 2.30 ft (0.7 m).

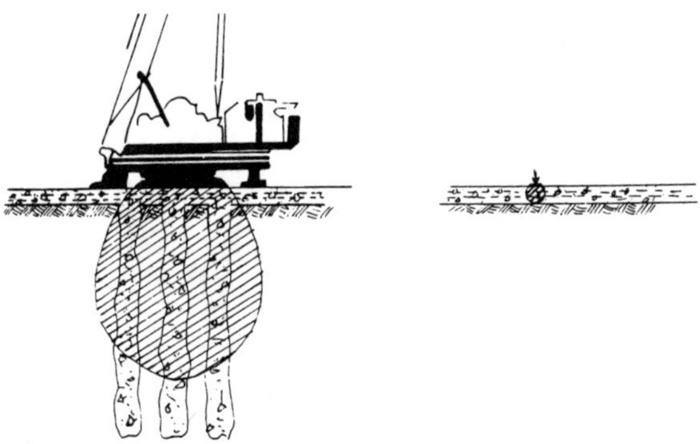

Fig. 8. Comparison of the bulbs of influence

Tables 1 and 2 present the k-values as measured during the sucessive stages of the test programme in three different locations, cf. Fig. 1a. The average values $\overline{k}$ are also given.

TABLE 1 : k-Value in Virgin Soil
($1\ N/cm^3 = 3.68\ lb/in^3$, 1 m = 3.28 ft)

	Virgin Soil		after placement of 0.4 m fill	
Location	k (N/cm^3)	$\bar{k}$ (N/cm^3) (1)	k (N/cm^3)	$\bar{k}$ (N/cm^3)
A	5.71, 5.21	5.46	5.40, 5.38	5.25
B	5.31, 5.31 4.52, 4.32	4.87	5.13, 5.13 5.71	5.32
C	4.27, 5.00 5.67, 5.59	5.13	5.50, 5.60 5.60	

(1) average

TABLE 2 : k-Values after Installation of Stone Columns
($1\ N/cm^3 = 3.68\ lb/in^3$, 1 m = 3.28 ft)

	Fill 0.4 m		Fill 0.7 m	
Location	k	$\bar{k}$ (1)	k	$\bar{k}$ (1)
	N/cm^3	N/cm^3	N/cm^3	N/cm^3
A	5.38 5.26 8.64 5.38	6.03	7.57	7.57
B	7.66 6.72	7.19	8.21 6.29	7.25
C	5.50 6.48 6.48 6.26 7.73 7.89	6.72	-	-

(1) average

Comparison of test results

As previously mentioned, SGT allows one to monitor the increase in stiffnes of the overall soil mass achieved by the installation of the sand columns. Table 3 summarizes the average k-values for the different stages of the test programme.

These values are graphically represented in Fig. 9 as a function of the thickness of the fill and clearly show the influence of sand column installation as well as of the layer thickness.

TABLE 3 : Summary of SGT Measurements
($1\ N/cm^3 = 3.68\ lb/in^3$, 1 m = 3.28 ft)

	$\bar{k}$ values (N/cm^3)			
Location (1)	non-compacted soil (2)	non-compacted soil + 0.4 m (3)	stone columns 0.4 m fill (4)	stone columns 0.7 m fill (5)
A	5.46	5.32	6.38	7.92
B	4.87	5.55	7.54	7.60
C	5.13	5.25	-	-
Average value	5.15	5.37	7.07	7.76

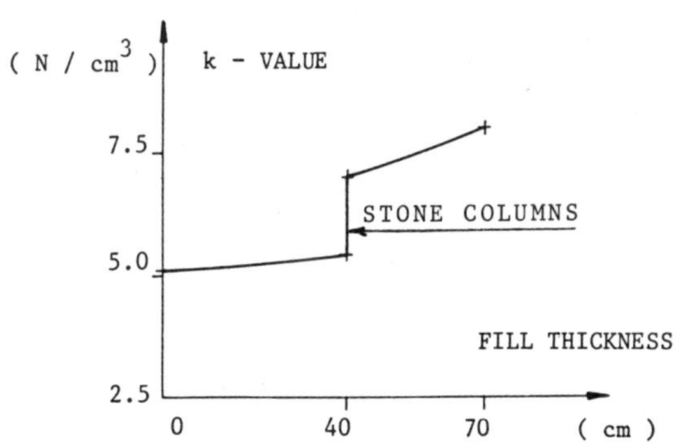

Fig. 9. Effect of reinforcement on measured k-value.
(1 cm = 0.394 in, $1\ N/cm^3 = 3.68b/in^3$)

CONE PENETRATION TESTS (CPT)

CPT tests were performed in the virgin soil as well as after installation in a stone column cf. Fig. 10. CPT 4, performed in the center of a stone column confirms that the length of the columns was about 26 ft (8 m). The CPT performed in the clay a few days after installation of the stone columns did not show any discernable change.

Although a drained long term settlement parameter cannot be directly correlated to an undrained strenght parameter such as q_c, empirical correlations have been developed for overconsolidated clays. However, as will be shown below, the scatter is large and this approach is thus not recommended for design purposes.

CPT tests were used to calculate the settlement for contact pressures of 1.45 psi (10 kPa) and 3.63 psi (25 kPa), respectively. The coefficient of subgrade reaction was then determined within this stress range.

The vertical stress σ_z along the axis below a circular rigid plate in a homogeneous, isotropic, elastic, half-space, can be expressed by :

$$\sigma_z = \frac{p_{av}}{2} \frac{1 + 3\zeta^2}{(1 + \zeta^2)^2} \qquad (5)$$

where p_{av} is the average contact pressure and
ζ is $\frac{z}{a}$ with z the depth where the stress is evaluated and a the radius of the plate

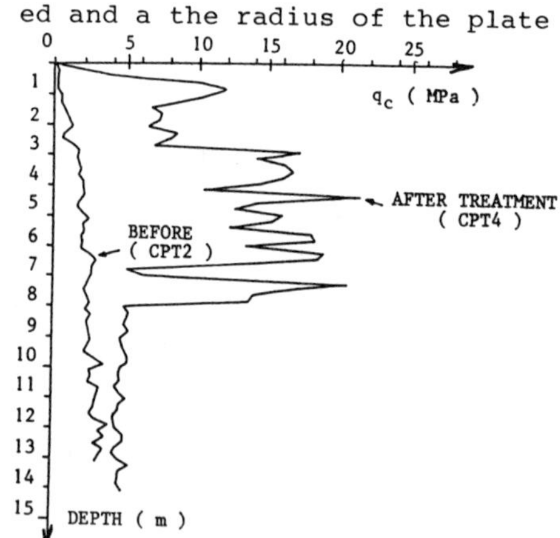

Fig. 10 CPT profiles before and in stone column, after treatment. (1 m = 3.28 and 1MPa = 6890 psi)

The settlements can then be estimated according to Terzaghi (5) :

$$\frac{\Delta h}{h} = \frac{1}{C} \ln \frac{p_o + \sigma_z}{p_o} \qquad (6)$$

where $\frac{\Delta h}{h}$ is the relative settlement

p_o is the geostatic stress at the depth considered

σ_z is the incremental stress due to the applied pressure

C is the compressibility constant which can be expressed by Cassan, (1)

$$C = \alpha \frac{q_c}{p_o} \qquad (7)$$

with q_c : cone resistance measured by CPT.

α : empirical coefficient depending on soil type.

The coefficient of subgrade reaction can then be calculated from the following ratio

$$k = \frac{p_1 - p_2}{\Delta h_1 - \Delta h_2} \qquad (8)$$

where p_1 and p_2 are the average stress of 1.45 psi (10 kPa) and 3.63 psi (25 kPa) respectively, acting on the central plate,

Δh_1 and Δh_2 are the respective settlements, resulting from the action of these two stresses

This k-value can be compared with the secant modulus determined from SGT measurements (cf. equations (1), (2), (3) and (4)).

Table 4 gives the k-values determined from CPT according to the above described procedure. As could be expected CPT3 did not show any increase in the cone resistance of the clay between stone columns. Table 4 also gives the k-values in the center of a sand column.

PRESSUREMETER TESTS (PMT)

Two pressuremeter tests were performed on the site in the clay before sand column installation as well as at the center of a sand column after treatment, respectivelly. The PMT results are summarized in Table 5.

TABLE 4 : Estimate of k-Values from CPT Tests
1 kPa = 0.145 psi, 1 N/cm³ = 3.68 lb/in³,
1 cm = 0.393 in)

			p_1 (kPa)	p_2 (kPa)	Δh_1 (cm)	Δh_2 (cm)	k (N/cm³)
before	CPT 1	$\alpha = 3$	10	25	0.428	1.170	2.03
		$\alpha = 4$	10	25	0.321	0.878	2.71
installation	CPT 2	$\alpha = 3$	10	25	0.920	2.100	1.27
		$\alpha = 4$	10	25	0.690	1.580	1.69
after installation	(1) CPT 4	$\alpha = 1.5$	10	25	0.129	0.297	8.93

(1) Test in stone column.

TABLE 5 : PMT in Virgin Soil and in a Column

		Virgin Soil		In Sand Column		
z (m)	E (kPa)	p_ℓ (kPa)	E/p_ℓ	E (kPa)	p_ℓ (kPa)	E/p_ℓ
1.0	3100	270	11.5	8500	960	8.85
2.0	5700	680	8.4	5700	710	8.03
3.0	5500	540	10.2	19100	1100	17.36
4.0	8000	800	10	9600	1100	8.72
5.0	15400	860	17.9	8700	1100	7.90
6.0	18000	970	18.6	8600	1100	7.92
7.0	14600	1100	13.3	14800	1100	13.45
8.0	13700	1100	12.5	14100	950	14.80

Following the recommandations suggested by Ménard (3) the settlement of the circular central plate can be calculated under a vertical stress, which yields the following estimates of the coefficient of subgrade reaction:

non compacted soil : k = 5.71 N/cm³ (21.0 lb/in³)
compacted stone column : k = 23.31 N/cm³ (87.8 lb/in³)

PLATE LOAD TEST (PLT)

A conventional plate load test (PLT) was also performed on the surface of the sand layer. The area of the plate was

750 cm^2 (11508 sq in). Since the depth affected by the test is on the order of 70 cm (27.5 in), only the sand layer has been actually tested.

Table 6 gives the plate settlements s for different stress levels p and the k-values determined at each stress level.

TABLE 6. Results of Conventional Plate Load Test ($1\ N/cm^3 = 3.68\ lb/in^3$, 1 mm = 0.039 in)

p (kPa)	s (mm)	k_{750} N/cm^3
50	0.99	50.5
100	2.86	35.0
150	4.69	32.0
200	6.54	30.6
250	8.33	30.1
300	10.00	30.0
350	11.86	29.5
400	13.53	29.6
450	15.05	29.9

COMPARISON OF TEST RESULTS

The average k-values obtained from SGT, CPT and PMT methods for the non compacted soil are :

$\bar{k} = 5.15\ N/cm^3$ ($18.95\ lb/in^3$) (SGT)

$\bar{k} = 2.37\ N/cm^3$ ($8.72\ lb/in^3$) (CPT)

$\bar{k} = 5.71\ N/cm^3$ ($21.01\ lb/in^3$) (PMT)

The obtained values are thus within the same range and globally of the order of $k = 5\ N/cm^3$ ($18.4\ lb/in^3$). This value corresponds to an E_s value of about 14 MPa (96.56 kpsi), which is in good agreement with the values (5-20 MPa 34.5 - 137.8 kpsi) in the literature for the Boom clay (6).

For the compacted soil, the results of the conventional plate load test can be used to assess the k-value of the larger plate load test, using Terzaghi's (5) recommended scaling law for sands :

$$k = k_1 \left(\frac{B+1}{2B}\right) \tag{9}$$

with k_1 : coefficient of subgrade reaction, determined for a 1 ft diameter plate

B : actual plate diameter in feet

CPT, PMT and PLT performed in the sand column give the following k-values for the large plate (k_{75000}) and for the 750 cm² (115.8 sq in) plate (k_{750}), respectively.

TABLE 7 : k-Values for Compaction Stone Column (1 N/cm³ = 3.68 lb/in³)

	k_{75000} (N/cm³)	k_{750} (N/cm³)
CPT	13.9	-
PMT	23.3	10.1
PLT	≃ 10	30

In spite of the variation of individual values, it can be concluded that a k-value of 20 N/cm³ (73.6 lb/in³) is representative for the 7.5 m² (78 ft³) large central plate of the Franki rig, resting on the compacted sand.

CONCLUSIONS

By instrumenting a Franki piling rig it is possible to determine in-situ the coefficient of subgrade reaction below a 7.5 m² (78 ft²) plate. The modulus values estimated from CPT and PMT tests agreed well with those of the large scale plate load test in the stiff to very stiff sandy clay. The k-value for the large plate was on the order of 5 N/cm³ (18.4 lb/in³), which is also in good agreement with the extensive data published.

SGT tests performed after installation of the displacement-type sand columns indicate that the k-value can be increased to 8 N/cm³(29.4 lb/in³), which corresponds to an increase of 60 %.

The test performed in the stone columns and in the fill material indicate that the k-value of the 7.5 m² plate resting on a uniformly compacted fill is on the order of 20 N/cm³ (73.6 lb/in³).

ACKNOWLEDGEMENTS

The concept of large-scale plate load tests using a

Franki piling rig was initiated by Prof. B. Broms who also made valuable suggestions during the implementation of the project. His assistance is gratefully acknowledged.

REFERENCES

1. CASSAN M (1978), Les essais in situ en mécanique des sols, Vol. 1 et 2 - Editions Eyrolles, 1978.

2. HOLEYMAN A. and WALLAYS M. (1984), Compactage par damage en profondeur, Proceedings of International Conference on In-situ Soil and Rock Reinforcement, Paris, October 1984, pp. 367-372.

3. MENARD L. (1967), Règles d'utilisation des techniques pressiométriques et d'exploitations des résultats obtenus pour le calcul des fondations. Notice D 60/67, Oct. 1967.

4. POULOS H.G. and DAVIS E.H., (1974), Elastic solutions for soil and rock mechanics, John Wiley and Sons, 1974.

5. TERZAGHI K. (1955), Evaluation of Coefficients of Subgrade Reaction, Géotechnique, V, 1955, pp. 297-326.

6. VAN BURM P. and MAERTENS J. (1977) Grondmechanische kaart a5.3.8 Antwerp, RIG and MOW, 57 p. + 9 maps.

BEHAVIOR OF SHALLOW INCLINED PLATE ANCHORS IN SAND

A.M. Hanna[1], B.M. Das[2], and A. Foriero[3]

Abstract

The ultimate pullout capacity of continuous shallow inclined anchors in sand has been investigated. Theoretically the concept of limit equilibrium has been employed, and the case of a two-dimensional failure mechanism has been considered. Laboratory model tests were conducted on instrumented model anchors which were tested at shallow depths with inclination angles varying from zero to 90° with respect to the vertical. Test results showed that the pullout capacity of these anchors increased due to an increase of the inclination angle and depth of embedment. The theory developed for anchor inclination angles varying from zero to 60° compares well with the present and available test results. Good agreement was achieved when the present test results for anchors inclined at 90° were compared with the theories given by Neely, Stuart, and Graham (1973) and Meyerhof (1973).

Introduction

There is increasing demand for soil anchors both on land and offshore to support structures subjected to tensile loading. In the design of transmission towers as well as deep-sea habitats, the requirement for uplift resistance is of particular importance. In the literature there have been numerous reports relating to the ultimate pullout capacity of horizontal (2,4,5,9,10,15,17,23) and vertical (3,7,19,20,21, 22) plate anchors. Some model test results on inclined plate anchors subjected to horizontal pull (8,14) and horizontal anchors subjected to inclined pull (15) have also been published. In contrast only a limited number of studies have been published relating to the ultimate pullout capacity of inclined anchors subjected to axial pull (1,11,13,

[1]Associate Professor, Department of Civil Engineering, Concordia University, Montreal, Quebec, Canada

[2]Professor and Chair, Department of Civil Engineering and Mechanics, Southern Illinois University at Carbondale, Carbondale, Illinois

[3]Graduate Student, Department of Civil Engineering, Concordia University, Montreal, Quebec, Canada

16,24). In practice, the design of these anchors is based on empirical procedure which could lead to either conservative or unsafe design. The present paper presents a rational approach which allows prediction of the ultimate axial pullout capacity of shallow continuous anchors embedded in sand and inclined anchors varying from zero to 60° with respect to the vertical (Figure 1). The theories developed by Neely, et al. (19) and Meyerhof (16) are recommended for the case of anchors inclined at 90° with respect to the vertical. Also the results of the experimental investigation on continuous model anchor plates in sand inclined at angles ranging from zero to 90° with respect to the vertical are presented. The present investigation is limited to shallow anchors where the embedment ratio D/H (where D=depth of embedment and H= height of anchor plate as shown in Figure 1) was equal to or less than 7.

Experimental Study

The sand used in this investigation is known commercially as Morie sand. The grain-size distribution and visual inspection classify it as clean, angular, uniform sand. Table 1 shows the summary of test results on the sand. The peak angle of shearing resistance (ϕ), as determined from direct shear tests at low stress levels, was 41.2° at a relative density of 63.33%.

In order to assure reproducible density, the sand was deposited in the testing tank from a fixed height through a metallic mesh. In order to conduct a test, sand was first deposited in the test tank up to a desired height. The anchor

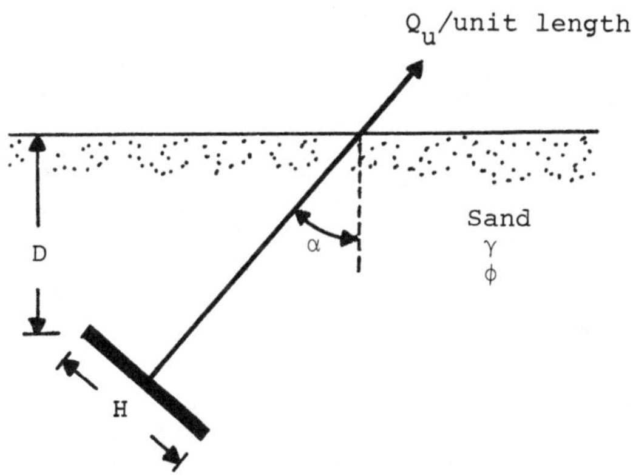

Figure 1. Continuous inclined plate anchor in sand subjected to axial pull

Table 1. Summary of Test Results on Sand

Parameter	Value
D_{10} (mm)	0.82
D_{60} (mm)	1.19
Uniformity coefficient	1.45
Specific gravity, G_s	2.66
Maximum void ratio, e_{max}	0.815
Minimum void ratio, e_{max}	0.590

was then placed in the desired orientation with the bottom of the anchor plate touching the top of the sand bed. Then sand was again deposited in the tank to bury the anchor to the desired embedment ratio.

The testing tank was designed to simulate the plane strain conditions of a continuous anchor. The inside dimensions of the tank were 152 mm wide, 1220 mm long, and 1015 mm high. The tank sides were made of plexiglass in order to observe the failure surface.

The reaction frame was designed for uplift loads up to 2.67 kN. The loading equipment included a motorized screw jack with a 900 mm maximum travel mounted on a steel plate bolted onto the steel tubing frame. The combination of a gear shaft, a gear reducer, and an electronic speed-controlled device provided the screw jack with a loading and unloading speed varying from 25.4 to 254 mm per minute.

The model anchor plates were made of an aluminum plate 12.5 mm thick. The plates were considered rigid enough for the expected loading. In order to simulate plane strain conditions, the model anchor width (B) at right angles to the cross section had to be the same as the testing tank width (152 mm). The heights (H) of the model anchors were 50.8, 101.6, and 152.4 mm. Pressure transducers were placed in circular openings cut in the plate, and the plates were covered with sandpaper. A hollow steel rod was fastened to the center of each plate. This rod served two functions, i.e., to pull the anchor plate and also as an enclosure for wiring. Figure 2 shows the schematic diagram of the experimental setup, and Figure 3a is a photograph of the actual setup.

For a given test, the pullout load was measured by means of a load cell connected at the top of the anchor rod. The pressures on predetermined points on the anchor plate were measured during testing by the system of transducers (Figures 3b and 3c). The difference between the loads measured by the load cell and the load resulting from the integration

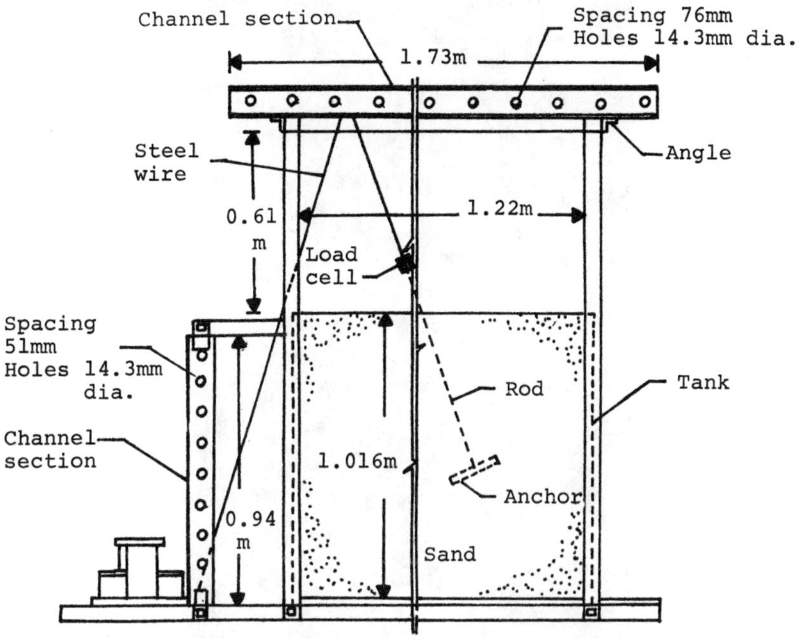

Figure 2. Schematic diagram of the experimental setup

(a)

Figure 3. Photographs showing (a) Experimental setup; (b) Model anchor (general view); (c) Model anchor (inside view)

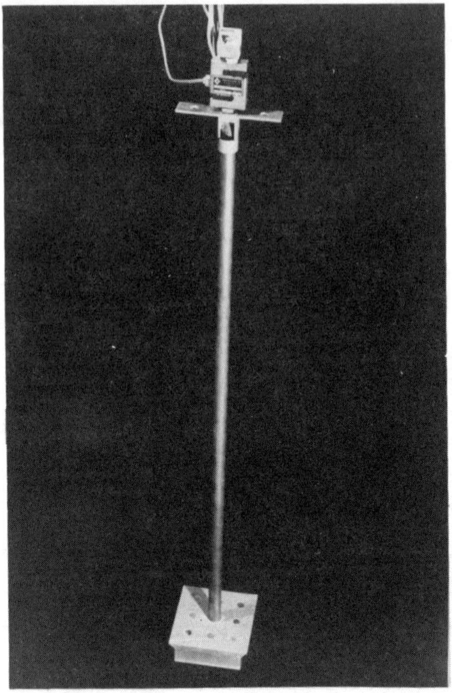

(b)

(c)

Figure 3. (Continued)

of the pressure diagram deduced from the transducer's readings revealed the friction load between the sand and the anchor in the failure wedge and the plexiglass sides of the testing tank.

Typical net pullout load (Q_u') vs. displacement curves are shown in Figure 4. Test results of the present investigation are summarized in Table 2, from which it can be seen that the ultimate pullout load increases with an increase of the angle of anchor inclination α and the embedment ratio D/H. Figure 5 shows a typical passive earth pressure distribution as measured by the transducers during testing.

Theoretical Developments

Observations during the experiments showed that, at failure, the anchor together with a soil mass of roughly truncated pyramidal shape was lifted up. These observations were noticeable for anchor inclination α varying between zero and 60°. For anchors inclined at 90°, the observed failure mechanism was similar to the classical ones proposed for developing earth pressure theories for vertical retaining walls. For anchors inclined at angles greater than 60° and less than 90°, the observed failure mechanism changes gradually between the two extremes (0 to 60° and 90°), respectively.

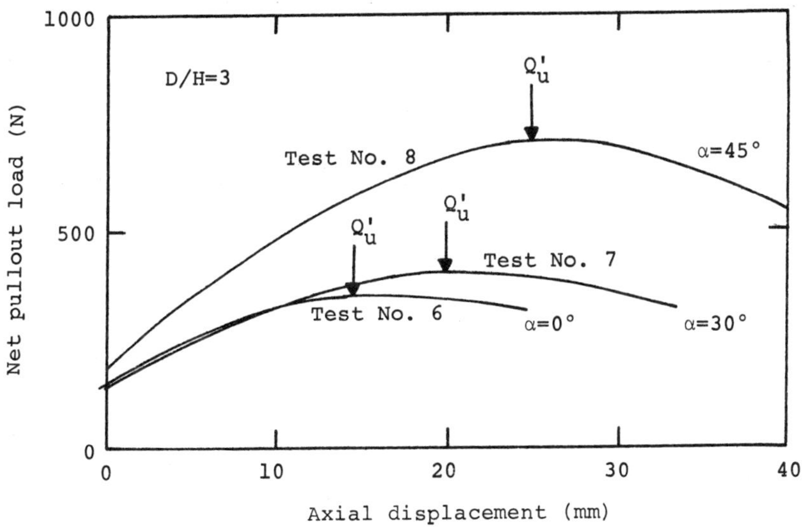

Figure 4. Plot of typical net pullout load vs. axial displacement

Table 2. Summary of Test Results

Test No.	Inclination angle, α (deg)	Depth, D (mm)	Plate height, H (mm)	Embedment ratio, D/H	Ultimate load: Load cell readings at failure, Q_u'(total) (N)	Ultimate load: Anchor plate load as integrated from transducer's readings, Q_u' (N)
1	0	304.8	152.4	2.0	255.52	193.89
2	30	304.8	152.4	2.0	289.52	221.83
3	45	304.8	152.4	2.0	483.24	390.94
4	60	304.8	152.4	2.0	839.98	749.51
5	90	304.8	152.4	2.0	1513.89	1411.76
6	0	457.2	152.4	3.0	403.53	332.41
7	30	457.2	152.4	3.0	456.08	381.81
8	45	457.2	152.4	3.0	771.36	681.52
9	60	457.2	152.4	3.0	1176.06	1080.18
10	90	457.2	152.4	3.0	2397.79	2290.86
11	0	609.6	152.4	4.0	630.16	554.02
12	30	609.6	152.4	4.0	787.07	707.33
13	45	609.6	152.4	4.0	1260.77	1163.45
14	60	609.6	152.4	4.0	2334.03	2131.06
15	0	330	101.6	3.25	235.56	178.87
16	30	330	101.6	3.25	261.12	200.42
17	45	330	101.6	3.25	440.72	369.42
18	60	330	101.6	3.25	783.99	699.58
19	90	330	101.6	3.25	1534.99	1442.07
20	0	482.5	101.6	4.75	392.13	325.25
21	30	482.5	101.6	4.75	461.95	390.27
22	45	482.5	101.6	4.75	760.91	672.36
23	60	482.5	101.6	4.75	1435.04	1334.24
24	0	634.9	101.6	6.25	570.61	499.69
25	30	634.9	101.6	6.25	687.92	617.64
26	45	634.9	101.6	6.25	1178.04	1100.79
27	60	634.9	101.6	6.25	2278.79	2180.09
28	0	355.7	50.8	7.0	207.88	158.99
29	30	355.7	50.8	7.0	250.72	196.30
30	45	355.7	50.8	7.0	373.72	306.82
31	60	355.7	50.8	7.0	742.34	666.72
32	90	355.7	50.8	7.0	1249.83	1165.52

In order to develop the theoretical model utilizing limit equilibrium analysis, the failure planes were assumed to be parallel to the anchor rod (direction of the pullout load) as shown in Figure 6. The assumption was made to simplify the analysis. The mobilized angle of shearing resistance on the assumed failure plane was taken as δ, while δ is less

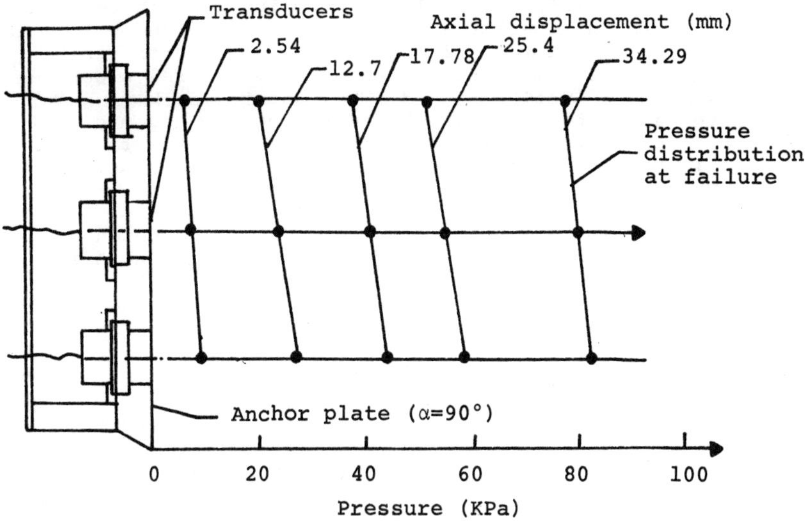

Figure 5. Typical passive earth pressure distribution behind an anchor plate ($\alpha=90°$; $D/H=3.0$; $\phi=41.2°$; and $\gamma=15.6$ kN/m^2)

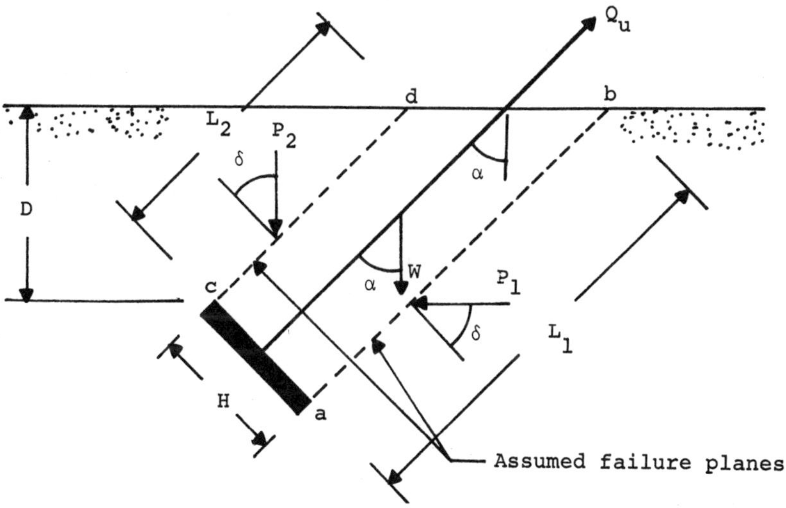

Figure 6. Assumed failure surface at ultimate load

than ϕ, which is the peak angle of shearing resistance of the sand. This failure mechanism is valid for an embedment ratio D/H (Figure 6) equal to or less than 7. At greater depths (D/H>7) the failure mechanism changes from general shear type (shallow anchor) to local shear type, and the anchors should be treated as deep anchors. The forces on the assumed failure surfaces *ab* and *cd* as shown in Figure 6 can be taken as the total passive earth force pressures P_1 and P_2 inclined at an average angle δ and acting downwards.

Thus for a continuous anchor plate of height H located at a depth D, the ultimate pullout capacity per unit length at right angles to the anchor section can approximately be given as

$$\frac{Q_u'}{B} = Q_u = P_1 \sin\delta + P_2 \sin\delta + W\cos\alpha \qquad (1)$$

where

$$P_1 = \frac{1}{2} R_\gamma K_{p\gamma} \gamma L_1^2 \qquad (2)$$

$$P_1 = \frac{1}{2} R_\gamma K_{p\gamma} \gamma L_2^2 \qquad (3)$$

where Q_u=ultimate pullout load per unit length of the anchor; B=length of anchor plate at right angles to the cross section shown (Figure 6); δ=average mobilized angles of shearing resistance on the assumed failure wedge; W=weight of soil block *abcd* and the self-weight of the anchor; γ=initial unit weight of sand; R_γ=reduction factor which depends on the ratio δ/ϕ (6); and $K_{p\gamma}$=coefficient of passive earth pressure for the weight component and corresponding to the angle of shearing resistance ϕ.

For determining the parameters given in Eq. (1):

1. If the analysis is made on the actual failure planes *ab'* and *cd'* (Figure 7a), then the mobilized angle of friction δ will be equal to ϕ. If the analysis is made on the assumed failure planes *ab* and *cd*, then the mobilized angle of friction δ must be less than ϕ, since failure has not taken place on the assumed planes (18).

2. The shear failure in the soil mass is a phenomenon of progressive rupture at various stress levels. Thus, while the slip line is progressing from point *a* to point *b'* (Figure 7a), the mobilized shear strength will vary from the peak value at point *b'* to a value less than the peak value on the rest of the slip line, and it may approach the residual value at point *a*. This behavior is reflected in the mobilized angle of friction δ on the assumed failure planes.

3. Observations during the experimental investigation

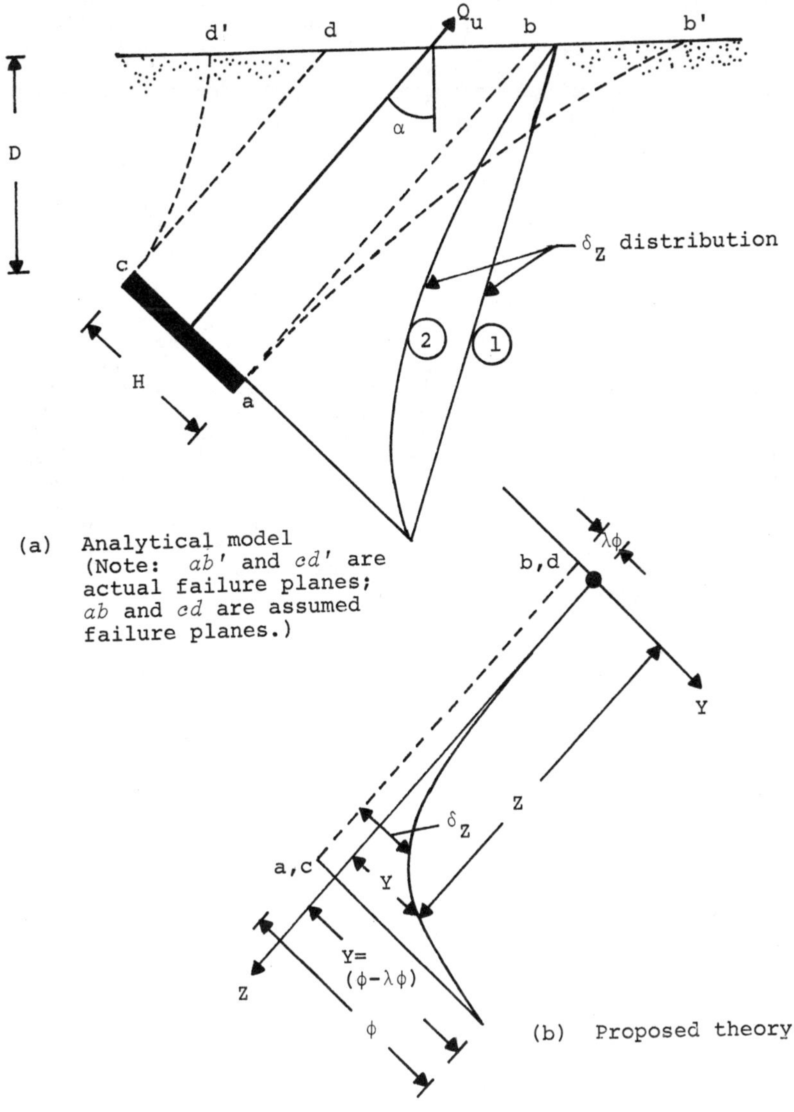

Figure 7. Distribution of the locally mobilized angle of shearing resistance δ_Z on the assumed failure planes

showed that, within the increase of the inclination angle α (within the range of zero to 60°), the assumed and actual failure planes approach each other. This will result in an increase of the angle of shearing resistance δ and, consequently, the developed passive earth pressure on the assumed failure planes (6).

A mathematical verification of these arguments is rather difficult. Also it is equally difficult to separate the effects explained herein in evaluating the average mobilized angle of shearing resistance δ and, consequently, the mobilized passive pressure $K_{p\gamma}$ on the assumed failure planes. However these difficulties may be circumvented by expressing the angle δ in a dimensionless form (δ/ϕ).

Based on the fact that the locally mobilized angle of shearing resistance δ_Z on the assumed failure plane at point *a* is equal to the angle of shearing resistance, it can be argued that moving upward from point *a* to point *b* on the assumed failure plane the angle δ_Z decreases to a minimum value of $\lambda\phi$ at point *b*. Note that λ is a coefficient which also depends on the inclination angle α of the anchor. Hence, the variation of the locally mobilized angle of shearing resistance δ_Z in terms of depth could be linear or nonlinear as depicted by curves 1 and 2, respectively (Figure 7a). Thus the total passive earth pressures P_1 and P_2 can be calculated from the following equations

$$P_1 = \int_0^{L_1} \gamma K_{(p\gamma)_Z} Z \cdot dZ \tag{4}$$

$$P_2 = \int_0^{L_1} \gamma K_{(p\gamma)_Z} Z \cdot dZ \tag{5}$$

The definition of Z is shown in Figure 7b. Combining Eqs. (2), (3), (4), and (5), the relationships for P_1 and P_2 can be given as

$$P_1+P_2 = \frac{1}{2} R_\gamma K_{p\gamma} \gamma (L_1^2+L_2^2)$$

$$= \gamma \left[\int_0^{L_1} K_{(p\gamma)_Z} Z \cdot dZ + \int_0^{L_2} K_{(p\gamma)_Z} Z \cdot dZ \right] \tag{6}$$

The left-hand side of the preceding equation can be evaluated by assuming an average ratio of δ/ϕ while the right-hand side can be computed utilizing an assumed distribution of the locally mobilized angle of shearing resistance δ_Z together with the coefficient λ. Solution of Eq. (6) was achieved by trial and error as explained below. A computer program was developed to perform such analysis.

1. Substituting Eqs. (2) and (3) into Eq. (1) and rearranging

$$\frac{Q_u - \frac{1}{2}(L_1+L_2)H\cdot\gamma\cos\alpha}{\frac{1}{2}\gamma(L_2^2+L_1^2)} = R_\gamma K_{p\gamma}\sin\delta \qquad (7)$$

2. The left-hand side of Eq. (7) is determined from the test data.

3. The right-hand side of Eq. (7) is evaluated by assuming an average value of δ/ϕ.

The solution proceeded by trial and error until the expression on the left-hand side of Eq. (7) was equal to the value of the expression on the right-hand side. Thus the experimental average values of δ, δ/ϕ, and R_γ were determined.

In order to predict the distribution of the locally mobilized angle of shearing resistance δ_Z on the assumed failure planes, it was assumed that (refer to Figure 7b)

$$Y = \frac{A'Z}{1-B'Z} \qquad (8)$$

with the origin at the soil interface where the assumed failure plane and the interface plane are assigned by the Y-Z axes, respectively, and constants A' and B' are given by

$$A' = \frac{\lambda\phi}{\beta} \qquad (9)$$

where $\lambda = f(\alpha,\phi)$

$$B' = \frac{1}{\beta} \qquad (10)$$

where β is a constant determined by boundary conditions, which means that along *ab*

$$\beta = \frac{L_1}{1-\lambda}$$

and along *cd*

$$\beta = \frac{L_2}{1-\lambda}$$

The preceding assumption is similar in nature to that which has been done for bearing capacity problems of shallow foundations on layered sand (12). Now, by substituting the experimental values of the average mobilized angle of shearing resistance into the left-hand side of Eq. (6) and using Eqs. (8), (9), and (10) in the right-hand side of Eq. (6), it was observed after several trials that λ can be approximated by the following equation

$$\lambda = \left(\frac{\alpha}{90}\right)^3 + e^{-5\tan\phi} \tag{11}$$

Design Charts

The computer program referred to earlier was used again utilizing Eqs. (8) to (11) which express the distribution of the locally mobilized angle of shearing resistance δ_Z and of the coefficient λ. The program was coded to determine the δ_Z distribution for a given value of ϕ. The purpose of this procedure was to determine the average mobilized angle of shearing resistance ϕ of the sand. The results of this analysis are presented in Figure 8. Setting

$$K_s \sin\phi = R_\gamma K_{p\gamma} \sin\delta \tag{12}$$

where K_s is defined as the uplift coefficient. Substitution into Eqs. (1), (2), and (3) leads to

$$Q_u = \frac{1}{2}\gamma K_s \sin\phi(L_1^2+L_2^2) + \frac{1}{2}(L_1+L_2)H\cdot\gamma\cos\alpha \tag{13}$$

Knowing the δ/ϕ ratio for given values of ϕ and α, the coefficient of passive pressure corresponding to the angle δ can be determined. Finally, the uplift coefficient can be calculated by using Eq. (12). For convenience, the variation of the uplift coefficient K_s was determined for the δ/ϕ ratios given in Figure 8. These results are given in Figure 9 in the form of design charts. Equation (13), together with the design charts, will allow the prediction of the ultimate pullout capacity of a strip anchor plate in sand. The present theory compared well with the experimental test results as well as other available test data of Wang and Wu (24) as shown in Figures 10a and 10b. However, the present theory underestimates the pullout capacity for anchors inclined at angles α greater than 60°. For anchors inclined at 90°, the theories developed by Neely, Stuart, and Graham (19) and Meyerhof (16) showed better agreement (Figure 11).

Conclusions

A theoretical and experimental investigation of the ultimate

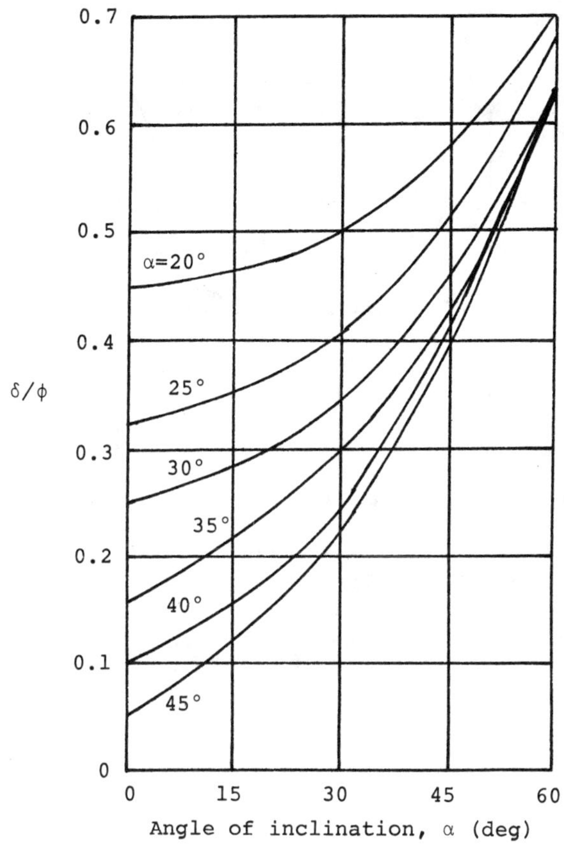

Figure 8. Theoretical values of average δ/ϕ ratios for strip anchor plates in sand

pullout capacity of shallow inclined strip anchor plates in sand have been conducted. The theory developed was verified with results from the present investigation as well as other investigations.

From the design charts, the uplift coefficient K_s can be determined from the known value of the angle of shearing resistance ϕ and the anchor inclination angle α with respect to the vertical. This angle of shearing resistance may be determined from the plane strain shear test results; however, for conservative design, direct shear test results may be used. The uplift coefficient can be used in conjunction with Eq. (13) to predict the ultimate pullout load of strip

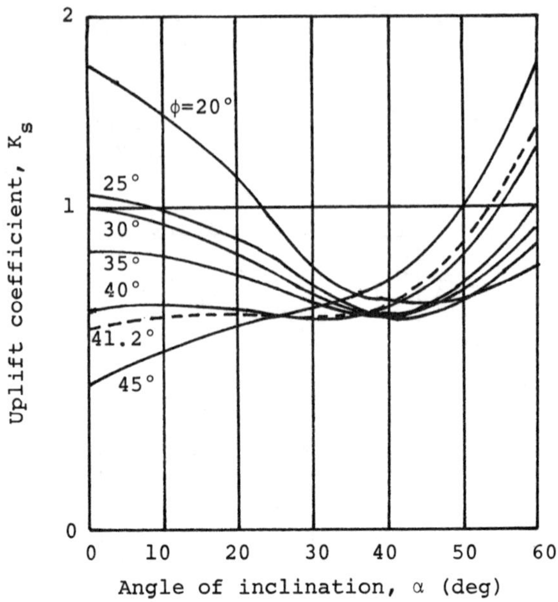

Figure 9. Design charts for determining uplift coefficient K_s

anchor plates in sand for the range of anchors inclined at angles varying from zero to 60° with respect to the vertical. For anchors inclined at 90° (i.e., vertical plates), the theories developed by Neely, Stuart, and Graham (19) and Meyerhof (16) are in good agreement with the present test results.

Acknowledgement

The financial support from the National Science and Engineering Research Council of Canada to the senior author is acknowledged. The experimental work presented in this paper was conducted by A. Foriero and M. Abumengal.

References

1. Abumengal, M., "Experimental Study on Shallow Strip Anchors in Sand," Master of Engineering Thesis, Concordia University, Montreal, Quebec, Canada, 1986.
2. Adams, J.I., and Klym, T.W., "A Study of Anchorages for

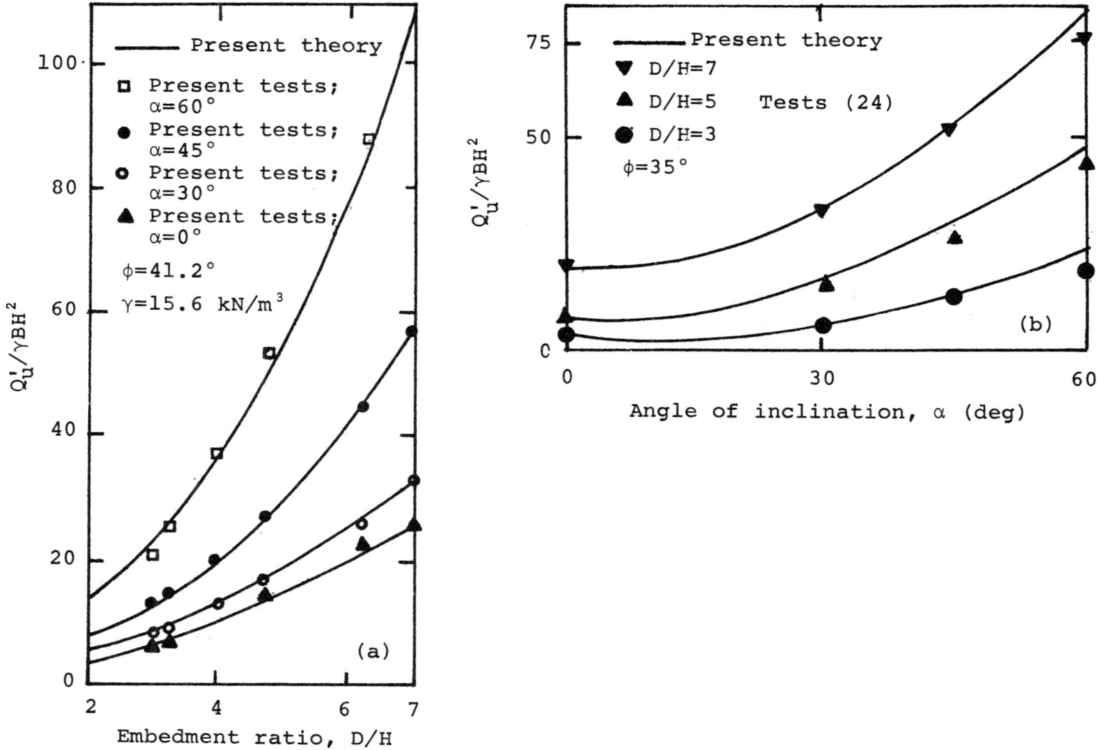

Figure 10. Comparison between the present theory and the experimental results.

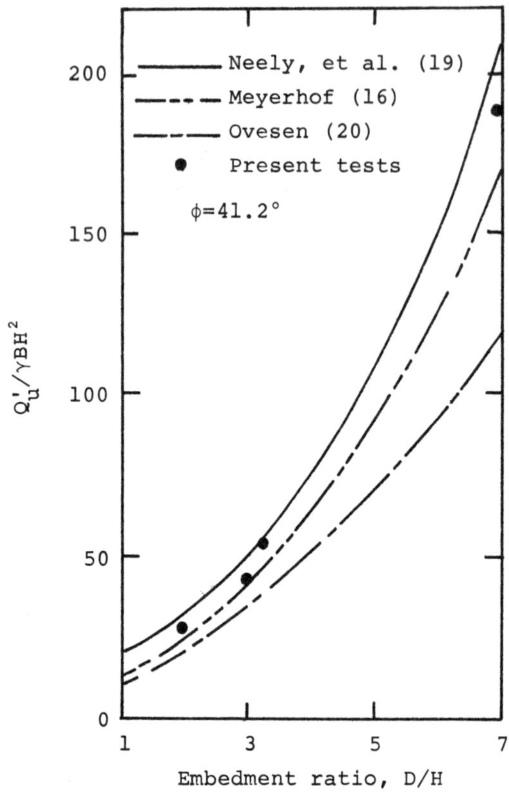

Figure 11. Comparison between the experimental results of the present investigation and other available theories (α=90°)

Transmission Tower Foundations," Canadian Geotechnical Journal, Vol. 9, 1972, pp. 89-95.

3. Akinmusuru, J.O., "Horizontally Loaded Vertical Plate Anchors in Sand," Journal of Geotechnical Engineering, ASCE, Vol. 104, No. GT2, 1978, pp. 283-286.
4. Balla, A., "The Resistance to Breaking Out of Mushroom Foundations for Pylons," Proceedings, V International Conference on Soil Mechanics and Foundation Engineering, Paris, Vol. 1, pp. 569-576.
5. Buchholz, W., "Erdwiderstand auf Aukenplatten," Jahrbuch Hafenbaulechnischen Gesellschaft, Vol. 12, PP. 300, 1930.
6. Caquot, A., and Kerisel, L., Tables de Poussee et Butee, Gauthier-Villars, Paris, 1948.
7. Das, B.M., "Pullout Resistance of Vertical Anchors,"

Journal of the Geotechnical Engineering Division, ASCE, Vol. 101, No. GT1, 1975, pp. 87-91.

8. Das, B.M., and Seeley, G.R., "Passive Resistance of Inclined Anchors in Sand," Journal of the Geotechnical Engineering Division, ASCE, Vol. 101, No. GT3, 1975, pp. 353-356.

9. Das, B.M., and Seeley, G.R., "Inclined Load Resistance of Anchors in Sand," Journal of the Geotechnical Engineering Division, ASCE, Vol. 101, No. GT9, 1975, pp. 995-998.

10. Das, B.M., and Seeley, G.R., "Breakout Resistance of Shallow Horizontal Anchors," Journal of the Geotechnical Engineering Division, ASCE, Vol. 101, No. GT9, 1975, pp. 999-1003.

11. Foriero, A., "Pullout Capacity of Inclined Strip Anchor Plates in Sand," Master of Engineering Thesis, Concordia University, Montreal, Quebec, Canada, 1985.

12. Hanna, A.M., "Foundation on Strong Sand Overlying Weak Sand," Journal of the Geotechnical Engineering Division, ASCE, Vol. 107, No. GT7, 1981, pp. 915-927.

13. Harvey, R.C., and Burley, E., "Behavior of Shallow Inclined Anchorages in Cohesionless Sand," Ground Engineering, London, Vol. 6, No. 5, 1973, pp. 48-55.

14. Hueckel, S., "Model Tests on Anchoring Capacity of Vertical and Inclined Planes," Proceedings, IV International Conference on Soil Mechanics and Foundation Engineering, London, Vol. 2, 1957, pp. 203-206.

15. Meyerhof, G.G., "The Uplift Capacity of Foundations Under Oblique Loads," Canadian Geotechnical Journal, Vol. 10, No. 1, 1973, pp. 64-70.

16. Meyerhof, G.G., "Uplift Resistance of Inclined Anchors and Piles," Proceedings, VIII International Conference on Soil Mechanics and Foundation Engineering, Moscow, USSR, Vol. 2, 1973, pp. 167-172.

17. Meyerhof, G.G., and Adams, J.I., "The Ultimate Uplift Capacity of Foundations," Canadian Geotechnical Journal, Vol. 5, No. 4, 1968, pp. 225-244.

18. Muhs, H., "On the Phonomenon of Progressive Rupture in Connection with the Failure Behaviour of Footings in Sand," Discussion of the Proceedings, VI International Confernce on Soil Mechanics and Foundation Engineering, Montreal, Canada, Vol. 3, 1965, pp. 419-421.

19. Neely, W.J., Stuart, J.G., and Graham, J., "Failure Loads of Vertical Anchor Plates in Sand," Journal of the Soil Mechanics and Foundations Division, ASCE, Vol. 99, No. SM9, 1973, pp. 669-685.

20. Ovesen, N.K., "Anchor Slabs Calculation Methods and Model Tests," Bulletin No. 16, Danish Geotechnical Institute, Copenhagen, Denmark, 1974.

21. Ranjan, G., and Kayshal, Y.P., "Load Deformation Characteristics of Model Anchors under Horizontal Pull of Sand," Geotechnical Engineering, Bangkok, Thailand, Vol. 8, 1977, pp. 68-78.

22. Rowe, R.K., and Davis, E.H., "The Behavior of Anchor Plates in Sand," Geotechnique, Vol. 32, No. 1, 1982,

pp. 25.41.

23. Vesic, A.S., "Breakout Resistance of Objects Embedded in Ocean Bottom," Journal of the Soil Mechanics and Foundations Division, ASCE, Vol. 97, No. SM9, 1971, pp. 1183-1205.

24. Wang, M.C., and Wu, A.H., "Yielding Load of Anchor in Sand," Proceedings, Symposium on Limit Equilibrium, Plasticity, and Generalized Stress Applications in Geotechnical Engineering, ASCE, Hollywood, Florida, 1981, pp. 291-307.

SOIL TREATMENT FOR FOUNDATIONS ON EXPANSIVE CLAYS

by

J. R. Blacklock* and A. D. Pengelly**

ABSTRACT

The recent growth in size and number of buildings constructed on expansive clay has generated field performance data with different foundation design technologies. One method of design and construction for the typical large one and two story building is a reinforced concrete stiffened slab supported on stabilized soil with the walls and prime structural components supported on drilled piers.

This paper documents the development of an engineering design and analysis method utilized for predicting post-injection total probable swell (TPS) based on both laboratory and field swell testing for a new stabilization technique that utilizes multiple injections, very close spacings, and combinations of lime slurry, lime/fly ash slurry and lime supernate to achieve necessary reductions in soil swell.

INTRODUCTION

The construction of large one and two story buildings on expansive clay soils offers a unique challenge to foundation design engineers in areas around the world. Most engineers would prefer to recommend a suspended structural floor supported on beams and piers; however, the economics of this type of construction often make it too expensive. This has caused engineers and owners to develop alternative designs utilizing stabilization technologies for soil treatment to reduce expansive clay swelling under the floor slabs, walks and parking lots prior to construction. One method of design and construction for the typical large two story building is a reinforced concrete stiffened slab on stabilized soil with the walls and prime structural components supported on deep drilled piers.

A new construction technology for soil stabilization which has recently been developed, tested and implemented on a project in the Dallas/Forth Worth area, utilizes a new injection method of soil stabilization. This paper documents the development of the new engineering design and analysis method utilized for predicting post-injection total probable swell (TPS) based on both laboratory and

---

* Professor of Construction Engineering Technology, University of Arkansas at Little Rock, Little Rock, Arkansas, Member ASCE.

** Construction Engineering Technologist, Woodbine Corporation, Fort Worth, Texas.

field swell test-ing. The new injection method utilizes multiple injection passes, very close spacings, and appropriate combinations of lime slurry, lime/fly ash slurry and lime supernate to achieve a reduction in soil swell. Multiple post-injection pressure swell laboratory tests are performed on soil samples from the injected areas to verify that the treatment method significantly reduces site differential swell as well as total swell of all treated soils.

Data from laboratory tests performed in conjunction with the 1986 injection of a full scale test pad and the analytical method for calculation of the predicted surface swell reduction, which utilized pressure swell tests on post-injection field samples are presented.

The entire volume of research associated with the building of foundations on expansive clay during the past 25 years has been very large and these efforts have been widely published and discussed. Contained within this volume of knowledge and literature is the state-of-the-art of lime slurry pressure injection (LSPI). Listed in the reference section of this paper are many of the important technical documents that have reported laboratory and field research projects and case studies of LSPI treatment of expansive clay soils. The first such documented efforts were in the early 1960's approximately 25 years ago, however, according to one Dallas, Texas geotechnical engineer, Haston (1981), "The process has a highly variable level of acceptance among design professionals," and "LSPI has been utilized on many projects to solve problems for which it is not suited." That all design professionals are still unable to understand fully the treatment benefits of LSPI, is therefore a fair statement. However, during the years since Haston wrote his technical paper there has been introduced by the injection industry an entirely "new look" LSPI process for treatment of expansive clay soils that answers most if not all of the noted short comings of the process detected during the first 20 years of research and development and still evident in 1981. The new look system called STABIL-PAD by the developing contractor to differentiate it from ordinary LSPI, contains many long sought after features including the following:

1.) Pre injection evaluation test procedures
2.) Post injection acceptance test procedures
3.) Guaranteed post injection results
4.) Rational method for foundation design
5.) Multiple injection procedures, and
6.) Site specific construction procedures

By reading the current (1987) literature data base it can be seen that the addition of the six above features to LSPI technology answers one by one most of the expressed concerns of past studies. Twenty years ago it was suggested by Higgins (1965 and 1969) that injections should be spaced much closer, and almost with out exception all succeeding researchers have arrived at the same conclusion. Still, as simple as this change sounds it was not easily achieved. Because of economic considerations, no one seemed willing to write a job specification that would actually meet this need for the customer. Haston (1981) introduced his technical paper with the following statement, "The design professional who seeks an inexpensive solution to the problem of foundations on expansive soil may choose to select LSPI as a means of reducing anticipated soil movements to levels which can be tolerated by the proposed structure." "The major benefits of LSPI are low cost and a reduction in expansive soil movements. No other commonly used measure of reducing expansive soil problems, especially in deep deposits of high expansive soil, is as low in cost." The key words in these sentences were inexpensive and low cost. Haston listed the 1981 Dallas foundation prices as follows:

1) LSPI - $0.35-$1.00 per sq.ft.
2) Placement of Select Fill - $1.50-$3.00 per sq.ft.
3) Structurally Supported Floor - $3.00 - $5.00 per sq.ft.

After 1981, in the middle of the recent Dallas building construction boom, research and development was originated to search for new improvements for the LSPI process, with the goal of designing a system of pressure injection stabilization for expansive clays that would include in the system a method for rational design and post-injection acceptance. This research and development effort, which was funded by a Dallas/Fort Worth based injection contractor, was begun by the authors and others in 1985 and was concluded in 1986 with the stabilization of the test pad documented in this paper. In the process, a new pre-jection swell test was developed and documented, Boyton and Blacklock (1985), and a new rational design concept was developed and is documented herein. Also a new closer spacing, multiple injection process and new post-injection acceptance test procedures were developed.

The purpose of this paper is to provide a description of the new look LSPI process and to present injection, test and analysis procedures, associated with its use, and to document the first actual full scale field test pad demonstration project.

BACKGROUND

The desire to economically treat expansive clay soils to prevent foundation failures has been the prime motivational force behind the development of the lime injection industry. Approximately 25 years ago lime slurry pressure injection (LSPI) as it is known today had a healthy beginning. Interest in lime injection in the United States was spread from the east coast to the west coast with full scale housing research projects in such diverse locations as Dallas, Texas, Jackson, Mississippi and Tucson, Arizona Building Research Advisory Board (1969). Several highway research projects were initiated and geotechnical engineers in the United States and Australia were performing research in support of the new LSPI technology.

Houses built in the expansive clay geological soil formations in the Dallas/Fort Worth area and financed by FHA in the mid-1960's were recommended to have the soils lime injected as a satisfactory alternative to more costly structural foundation systems. In short, lime injecton seemed to be rapidly becoming the prefered system for treating expansive clay soils. It was supposed by many that lime seams, though placed by widely spaced injections, would grow together and through a theory call "lime migration" the entire soil mass would be stabilized. But, research began to show that in fact, lime does not migrate more than 0.50 - 2.0 inches in fine grained soils and it did not take long for the word to be passed around that lime did not migrate. The bubble had burst and lime injection could have died, but it didn't. Hope stayed alive among some engineers. A few small contractors continued to sell lime injection services for soil stabilization. More research was then funded by FHA, FRA and others in the 1970's Blacklock (1977) and Thompson (1975). As engineers begin to learn more about lime injection, its use grew, especially where the expansive clay problems were severe and alternative foundation solutions were costly.

It was realized by engineers and contractors that with lime injection, more was better, so by the 1980's most stabilization projects were double injected and many jobs also specified water injections to pre-swell the site. But still no real progress was made with pre-injection evaluation testing or post-injection verification testing. The lime injection process used to treat expansive clays could still be described as more of an art than a science. This was true in spite of the considerable progress that had been made in the strength technology of lime injection and also of the new procedure advanced for lime/fly ash injection. Laboratory testing to support strength injections was advancing nicely, but no accompanying progress was being made for swell testing technology. Even today engineers are writing in technical reports that lime injection will reduce swell in soils 50%, which is at best a guess. They seem to be saying, "We know it works, but we don't know how it works or how much it will reduce swell so a good guess is 50%."

In 1985 it became obvious to some that new procedures were essential for engineered injection stabilization of swelling clay soils. The outgrowth of this desire produced new thoughts about injection and new laboratory test procedures were advanced for both pre-injection evaluation and post-injection verification. Construction procedures were modified drastically, injection spacing was changed to a 2-ft grid or smaller and injection specifications were designed to be site specific. For the first time in the history of injection, geotechnical engineers were able to perform post-injection swell tests and verify that the swell had been reduced by injections. The new injection system for treating swelling clays appears to offer promise. In many ways, it has helped answer the laboratory and reserch questions of the past 25 years. Now it would be good to proceed once again with the research intent of the early 1960's, to study the new multiple lime injection process and search for true engineering understanding. Then in a few years, engineers would no longer have occasion to write that lime injection is confusing and contraversial.

PAST RESEARCH

During the past 25 years that LSPI has been under development, much has been written and many research projects have been conducted in an effort to understand the LSPI process. Many of these have been listed in the References at the conclusion of this paper and several of these have been abstracted below. Space does not allow enough to be written about each to fully expand on their scope, but it is felt that the information provided will add to the general feeling of understanding of the subject for those reading this paper. An emphasis has been placed on expansive soil applications in general and some specific projects in particular. Acknowledgement and thanks is given to each whose paper has been used and apology to those that have been omitted.

In 1963, an engineering firm was asked to investigate in detail the clay subgrade below the proposed Dallas Apparel Mart in order to develop construction techniques to stabilize the clays and permit the construction of a 250,000 square foot concrete slab floor directly on the ground surface. In their project research report, Haynes and Mason (1969), state, "It was concluded that the most efficient system of prewetting the subgrade soils was by the use of water mixed with the surfactant - - -." "In order to reduce the secondary swell phenomenon and also to reduce the possibility of future shrinkage of the clay, it is believed the use of lime in connection with prewetting of the soil was economically justified." Their research concluded that although prewetting treated primary swell that long term secondary swell was reduced by the use of lime as a secondary retarding agent.

Thompson (1965), published a noteworthy article that discussed in detail the improvements in engineering properties of lime soil mixtures and the four basic reactions - cation exchange, flocculation and agglomeration, carbonation, and a pozzolanic reaction. He concluded the article by stating, "The lime-soil reactions that produce improved plasticity, workability, shrinkage and swell properties readily occur with all fine-grained soils and are not dependent on natural soil properties." "Strength producing lime-soil pozzolanic reactions proceed to varying degrees with different soils and the reaction is significantly influenced by such soil properties as organic content, pH, soil weathering and soil drainage."

During the fall of 1965 a soft glacial clayey silt deposit was pressure-injected with lime slurry to depths of 20 feet to determine the effects of inplace lime stabilization for a highway foundation in Pennsylvania. The research report by Lundy and Greenfield (1968) stated, "High pressure lime slurry injection was moderately successful in forcing a slurry uniformly into the soft clayey silt deposit - - -." "The results of the project revealed a definite usefulness for high pressure lime slurry method of in-situ stabilization in cohesive soils."

In the early 1960's, Louisiana Department of Highways initiated a comprehensive study to test the effectiveness of in-place lime treatments. One method evaluated was the new high pressure lime injection method. In interim report No. 2, Higgins (1965), stated, "Results obtained indicate that a good distribution of lime was achieved at depths up to twenty feet." In the final report 1969, Higgins (1969), stated, "The results of this study indicate that of the three methods tested, the high-pressure method is the most effective." "With refinements and improvements, such as closer injection spacing or perhaps continuous injection, the high-pressure injection process might become a workable and useful tool."

Pressure injection of lime slurry was explained in detail and offered as an experimental in-situ stabilization treatment for expansive soils in the publication "Chemical Soil Stabilization" Building Research Advisory Board (1969). In discussing deeper treatment methods, the report stated, "Of the three lime methods for deeper treatment, pressure injection of lime slurry is the most widely used. Its principal advantage over drill-lime is that considerably more lime can be used and the slurry is immediately forced outward to a distance of several feet. The slurry follows the paths of least resistance; e.g., moving along cracks, bedding plane, root lines, or passageways created by the jetting slurry. The applied pressure generally exceeds 200 psi. Admittedly, the distribution is not uniform, with the lime solids being deposited in a network of horizontal, sheet-like seams, often interconnected with vertical or angular veins. But by using close spacings, the network created is believed to be generally quite extensive."

"Following treatment, the ground water percolation helps in the calcium ion migration; thus, lime's stabilization effects are gradually extended outward from the seams, resulting in gradual improvement in the soil mass. However, this improvement is not manifested in significant PI reduction or strength gains. Rather the improvement is considered to be due to two factors: (1) the injection of so much water tends to pre-swell the soil, and (2) the sheet-like seams of lime serve as individual moisture barriers, which tend to stabilize the moisture content of the soil and eliminate serious moisture fluctuations that could cause disruptive swelling and shrinkage."

Ingles and Neil (1970), reported Australian research concerning the lime slurry penetration and moisture movements in soil in which they concluded the following:

1) "Fissuring of the soil and the amount of moisture associated with the lime slurry are major parameters for practical application of the lime slurry injection process."
2) "The penetration of lime slurry appears to be correlated only poorly with laboratory permeabilities, and lime slurry penetrated only fissures and not pores. Interconnection of the fissure system is therefore important to the design of the lime slurry injection process."
3) "The presence and extent of groutable fissures can only be determined by a field test. Calculation of penetration by formulae requires soil profile parameters measuring effective hydraulic radius, permeability, or porosity. For cracking clays none of these properties can satisfactorily be measured."
4) "The experimental field program indicated a qualified effectiveness, as regards volume stability and stiffness of the soils for lime slurry injection, the qualification being that

expansive soils are most advantageously treated at times of maximum desiccation."

Fohs and Kinter (1972) reported the results of a laboratory study to obtain information on the movement (migration) of lime in compacted clay to measure the horizontal movement of lime, as this was thought to more closely represent field conditions where lime is injected or placed in drilled holes. Statically compacted blocks of each of the two fine clay soils were prepared at optimum moisture content and AASHO T 99 maximum dry density. In conclusion, they stated, "Lime migrated into the soil from the slurries in amounts ranging from 2.5 percent at a distance of 0.19 from the slurry-soil interface to about 0.8 at 1.5 to 2.0 inches. - - - From the data obtained for the experimental conditions employed in this study, it is concluded that migration of lime does take place when lime slurry is applied to compacted soil." But, they felt that the small distance of migration was of no consequence for stabilization with drilled lime and lime injection and further stated, "If any actual benefits have been obtained from such field applications as have been made of these two techniques, it is concluded that unknown factors other than lime migration must have been responsible."

Thompson and Robnett (1975) presented a paper with discussion of diffusion-migration of lime in which they stated, "(1) Intimate permeation of fine-grain soils through pressure injected lime is virtually impossible, (2) lime slurry can be pressure injected into certain fine-grained soil masses if varves, seams, fissures and cracks exist but the distribution of the slurry is stratified or of a network type. However, the lime will tend to be translocated in the fine-grained soil mass as time progresses because of diffusion and migration phenomena."

Haston (1981), described the LSPI state-of-the-art and included recommended job specifications and recommended laboratory testing and inspection services. In the paper he lists six limitations of LSPI. He stated, "They include a poor and short experience record, difficulty of verification, general confusion, lack of knowledge, and a history that includes many cases of improper use of LSPI process. - - - There is much confusion, and no general agreement, among designers over the proper procedures and specifications for LSPI. - - - No guarantee can be made that LSPI will in fact perform the intended reduction in potential soil movement. - - -There is a generally inadequate level of knowledge among design professionals concerning the effect of LSPI on soil."

McKeen (1981) reported the results of an FAA study into design of airport pavements for expansive soils in which he determined that measurement of soil suction was the key step in quantifying moisture-induced soil behavior and procedures were outlined for suction characterization of the soil and for estimating the in-situ differential movement likely to occur under pavement. Of the sites that he investigated with field and laboratory tests, one was the Dallas/Fort Worth regional airport where soil formations outcropping are the expansive Woodbine and Eagleford clays, both typical of soils throughout that region of Texas.

In another paper on the subject, McKeen and Hamberg (1981) stated that most expansive soils encountered in engineering problems are at a degree of saturation below 100 percent and that knowledge of the moisture condition of such soils is best obtained by measuring soil suction.

Blacklock (1982) presented information on the lime slurry glaze method of soil testing which was a new aproach for pre-injection evaluation of strength improvement of a lime injected soil mass. Strength data determined from compression testing of laboratory manufactured lime glaze (coated) test specimens were used to indicate the influence of injected lime seams on soil mass strength. The method development was later continued and its final documentation was published in the Lime Slurry Pressure Injection Bulletin, Boyton and Blacklock (1985). In addition to the lime glaze test presented in Appendix A of the Bulletin, a lime/fly ash seam stabilization test procedure was documented in Appendix B and the new lime seam swell test for pre-injection evaluations was presented in Appendix C. Copies of these test instructions can be obtained by writing the Nation Lime Association in Arlington, Virginia.

Petry, Armstrong and Chang (1982), reported the results of a research program to determine the short term active soil property changes caused by injection of lime and lime/fly ash slurries. The study involved a field experiment at the Dallas/Fort Worth airport where 12 areas were treated with four replications of three treatments and four areas were used as untreated controls. Those properties concluded to have significantly been affected by injection stabilization included water content, plastic index, swell potential, cation exchange capacity, calcium levels in pore water and calcium levels in the exchange complex. The ground level monitoring analysis supported the stabilizing effects of LSPI.

In two milestone papers which documented the enormous extent of damage caused by expansive clays to houses and buildings in the United States, Jones (1972) and Jones and Holtz (1973), the civil engineering community was alerted officially that the problem of expansive clay was an engineering problem and that the damage each year was in the range of billions of dollars, perhaps far larger than anyone until that time had imagined.

In a recent Civil Engineering article by Jones and Jones (1987), expansive soil damage in the United States is estimated to cost at least $1 billion a year to homes alone and perhaps as much as $9 billion a year to all built structures. They point out that soil-preparation options can be split into three groups: (1) reworking the soil; (2) controlling its moisture content, or (3) adding chemical stabilizers. Concerning LSPI, they state, "The technique provides good stabilizing results in some cases, and often is workable even though the clay is not reactive (does not gain strength) with lime or only moderately so, but sometimes has only limited long term effect. - - - Lime slurry pressure injection is reasonable for use in any area where it has a demonstrably effective track record. In such cases, it is the most economical chemical stabilization technique."

INJECTION TECHNOLOGY (State of the Art)

The injection of slurries and grouts for the purpose of stabilizing soil has progressed technologically as the practice has been continually implemented to solve a wider variety of soil problems. As more information has been generated about the nature of expansive clay soils, ideas concerning modifications to the injection methods used to treat the expansive clays have also developed. Along with modifying some long practiced techniques, new equipment capable of deeper penetrations has been placed into regular use. The slurry itself has also undergone changes, with new designs to achieve specific goals for each injection pass. All of the modifications to injection construction techniques, slurries, and equipment have enabled the contractors to provide a more predictable and reliable finished product.

The technique of injecting slurry is accomplished by a vehicle such as that shown in Figure 1. The slurry injection mechanical unit is normally mounted on the front of the injection vehicle. The most common injection mechanical unit is a hydraulic-lift mast with cross-beams which contain the four injection rods. In this type of unit the rods are forced into the ground simultaneously by the hydraulic action of the lift mast beams. A heavy duty larger mechanical injection unit is manufactured of four self contained injection towers with a hydraulic motor driven power system. This type of injection unit is used for dense clay stabilization and other projects which call for difficult penetration or deep applications. The injection rods are operated independently which gives additional control to the operator, as well as more force per rod. The slurry is usually mixed in a large batching tank that is typically 10-ft in diameter and 40-ft long and the slurry is then pumped with high pressure hoses from the slurry tank to the injection vehicle.

The most common stabilizing material used in slurries for injection stabilization of expansive clay soils is lime. The chemistry of lime is such that it reacts with minerals found in expansive clay to render the clay particles non-expansive. The dry lime to water ratio is typically specified at 2 1/2 -lb of dry lime per gal of water. The lime is delivered to the jobsite as hydrated lime or quicklime and is then unloaded from the delivery truck into the water in the slurry tank. Lime/fly ash is also slurried and injected into expansive clays. Lime/ /fly ash serves to form a moisture barrier especially in large cracks. In some cases a lime/fly ash slurry can be injected around the perimeter of a building pad to isolate the soil mass, which makes treatment of the soil with LSPI more effective. Various research investigations have shown that some of the benefits achieved from injection of lime slurry include, but are not limited to, a general increase in soil moisture content (3% - 5% per injection), a reduction in the observed range of moisture contents, an elevated pH, and a filling of the natural soil crack pattern, Figure 2, with lime slurry seams which stabilize the clay and help

Fig. 1 - Typical Injection Machine with 20-ft Towers

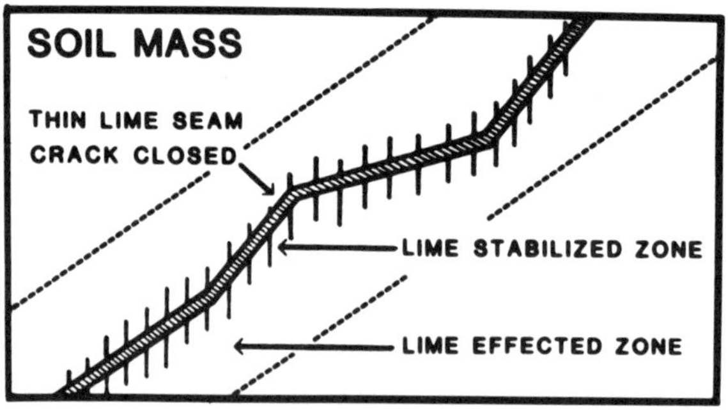

Fig. 2 - Soil Crack after Stabilization with LSPI

prevent the movement of moisture through the soil.

Even though the use of LSPI has proven itself both in the laboratory and in the field on many jobs, it still has not been standard practice because of its noted short comings and lack of a rational method of design. In the past, the use of soil moisture content criteria has been the main practice for acceptance of lime injection projects. Achieving soil moisture content equal to 1/2 the liquid limit has been the standard by which most injection work has been measured. A pocket penetrometer has also been used on extruded shelby tube samples with acceptance criteria being a reading of 3.0 TSF or less. A swell test which could actually predict a numerical amount of movement based on calculations from data collected from post-injection samples has not previously been utilized.

DEVELOPMENT OF STABIL-PAD CONCEPT

The concept of a highly improved injection system with measurable results became a reality after the development of new swell test method to predict the results of injection treatment in advance. To be able to guarantee an injection system, it would be necessary to know ahead of time how to specify the injection needed to meet a requirement of an inch or less movement in the treated zone. To test soils for their potential swell and their reaction to injection treatment the "Lime Slurry Seam Swell Test" was devised. The Lime Slurry Seam Swell Test uses remolded soil samples to creat a sample with a seam. The amount of material in the seam can be increased to simulate a greater number of injections. After curing, these samples are tested to determine swell reduction versus treatment.

The concept of injecting a site with slurries and testing the soil afterwards utilizing swell tests to predict the Total Probable Swell (TPS) required the development of an analytical method. And finally, in order to achieve the maximum amount of swell reduction the injection technology was modified from typical injection methods to a state-of-the-art application which was based on information derived from research conducted with seam swell samples. The STABIL-PAD system of building pad injection stabilization utilizes a maximum 2-ft spacing of the injections to insure maximum permeation of the clay soil. The depths of injection differ from a typical injection in that they are sequentially layered at various depths as opposed to one uniform depth. The slurries used consist primarily of lime slurry, lime supernate, and lime/fly ash. Lime slurry is the traditional slurry used for injection stabilization. Lime supernate is used as an alternative to increase moisture content of the subsurface soils. Lime supernate is a clear solution with enough lime present to raise the pH to approximately 13. The supernate has calcium cations available for ion exchange with the clay particles. Water has no such ability and therefore water only injections are not used. Lime/fly ash slurry is used for sealing large cracks and fissures through which free water could travel. Such a slurry is useful to seal off the perimeter of a building which will help retain moisture under the slab. The variety of slurries used allows for a more complete approach to the solutions of the soil swell problem.

ANALYTICAL METHOD

The analytical method for determining swell utilizes post-injection swell tests to calculate the Total Probable Swell (TPS) of the building site. The concept of TPS utilizes past research data concerning the swell potential and moisture movement potential of in situ clay soils. When the potential swell is caused to be limited by the potential moisture movement then that which occurs is termed the TPS. The TPS is a calculated total surface movement that is predicted to occur based on actual post-injection laboratory pressure swell tests and the analytical method.

The beginning of the analytical method dates back to 1973 when Davis and Tucker (1973) presented research findings concerning the soil moisture and temperature variations beneath a slab barrier on expansive clays. They observed that the soil in the active swell zone in expansive clay contributed less and less to the surface heave as a function of depth and that the prediction curve could be constructed from data showing the maximum range of moisture contents, Figure 3.

Based on this curve, Davis and Tucker predicted that in the Dallas/Fort Worth area, 85% of the swell typically occurs in the top seven feet. Following this concept, of reduced probable swell for deeper soils, they advanced the new terminology, seasonal change factor, F, with values from 1.0 at the surface to essentially zero for deeper soils, Figure 4. These values are calculated directly from the range of observed moisture contents. Using this theory, the analytical method was developed in 1986 to support the first STABIL-PAD project. The concept of reducing the swell to measurable post-injection values came first and the analytical method came second in an attempt to evaluate the results of the lime and lime/fly ash slurry injection by calculating the TPS.

It was decided to use post-injection swell data from borings located inbetween the injection holes as base line data and to calculate the TPS for the site according to the following equation DAS (1984):

$$TPS = [\ (Si)\ (Fi)\ (Hi)\ (0.01)]$$

Where Si = swell, in percent, for layer "i"
Fi = seasonal factor for layer "i"
Hi = thickness of layer "i"

DEMONSTRATION TEST PAD

The STABIL-PAD concept had evolved by early 1986 at approximately the same time that preliminary design of a new retail facility in Irving, Texas was taking place. The site of the proposed building was located directly north of state Highway 183 and approximately 1 mile west of Beltline Road in Irving, Texas. Preliminary investigations revealed unsuitable expansive soil throughout the site. As the various choices of foundation design were compared economically, the STABIL-PAD concept was the most attractive. However, since a site had not previously been injected to achieve a level of performance

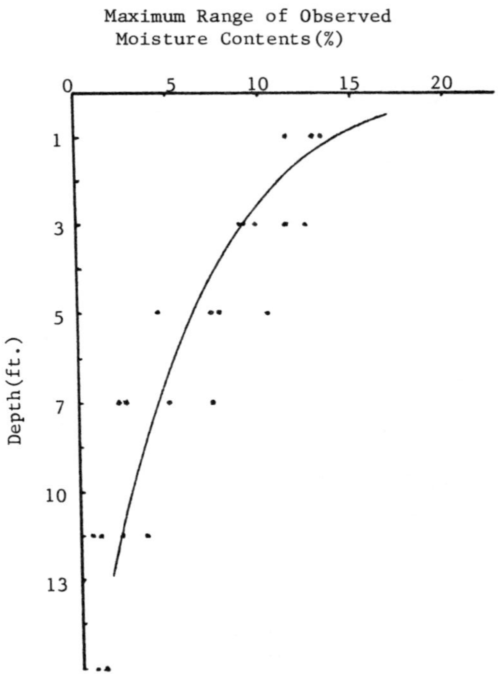

Fig. 3 - Maximum Range of Observed Moisture Contents from Field Borings

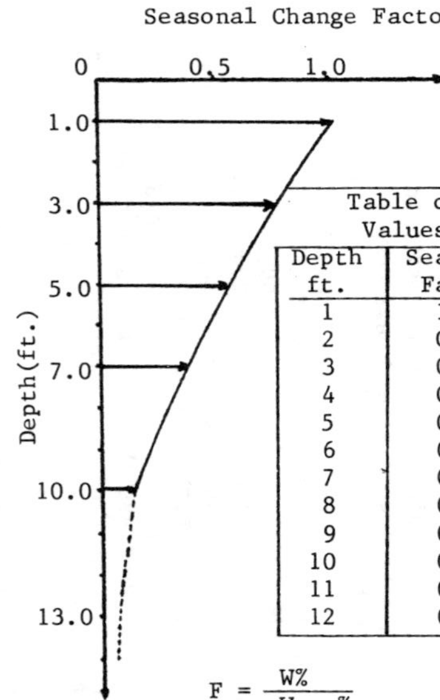

Table of Values

Depth ft.	Seasonal Factor
1	1.00
2	0.82
3	0.69
4	0.59
5	0.50
6	0.40
7	0.31
8	0.27
9	0.18
10	0.12
11	0.09
12	0.05

$$F = \frac{W\%}{W_{MAX}\%}$$

W% = Maximum Seasonal Moisture Change at Each Depth

Fig. 4 - Seasonal Change Factor vs. Depth

based on a post injection swell test, it was obvious that a performance type test section should be constructed and evaluated. The proposal to inject a test pad was made to the owners and the idea was accepted with a go or no-go provision. A 50-ft by 50-ft area was designated for treatment with the proposed STABIL-PAD system to prove it could achieve the foundation design engineer's criteria of performance.

INJECTION PLAN - The injection plan consisted of site specific details as to how the injection work was to be accomplished. Firstly, several layer depths were selected. The injections on the test pad were to be at 7-ft, 10-ft and 12-ft respectively, Figure 5. The stratigraphy of the site basically consisted of two layers. The upper layer thickness varied from 6-ft to 8-ft and consisted of a weathered dark brown clay. The lower layer was typically tan, weathered to unweathered shale, which was very expansive. By design, the majority of the injected material was to be placed in the upper 7-ft. The injection pattern was laid out on a 2-ft grid pattern, Figure 6. The plan was to achieve the maximum clay permeation possible. A total of six injection passes were made on the test pad over a 12 day period. A 48 hour curing period was allowed between each injection pass. After the completion of the first injection pass the free lime slurry on the surface of the pad was mixed and compacted to seal the surface and to provide a working platform for the remainder of the injection work.

TESTING - After the final injection pass was completed the injected soil was allowed to cure for five days before soil borings were made to obtain post-injection swell samples. The borings were taken in 5 locations, 3 on the pad and 2 in adjacent untreated areas. The nature of the soils in the treated versus untreated areas was somewhat different which caused difficulties in comparing the data. Several types of tests were run on the soil including moisture content determinations, Atterberg Limits, linear shrinkages, and pressure swell. Several interesting observations were seen from the test data. The data showed that soil changes as a result of STABIL-PAD injections caused the average pH to be increased 1.6 in the upper 7 feet, and 0.1 from 8 - 15 feet. The average Plasticity Index was decreased 4.0 points in the upper 7 feet and 3.1 points from 8 - 15 feet. These are thought to be significant changes in soil properties as a result of the injections and further substantiate the added benefits of multiple injections and close spacings.

ANALYSIS - The data from the swell tests were used to calculate cumulative swell in both the post-injection treated soils and control untreated soils. The post-injection swell test data is shown in Tables I and II. The average swell was taken for each layer of soil. The far right hand column shows the cumulative swell. Even without taking into consideration the seasonal factor "F", the TPS in the top 7-ft is still only 0.84 inches. Tables I and II do not reflect the calculation for swell reduction based on seasonal moisture variation.

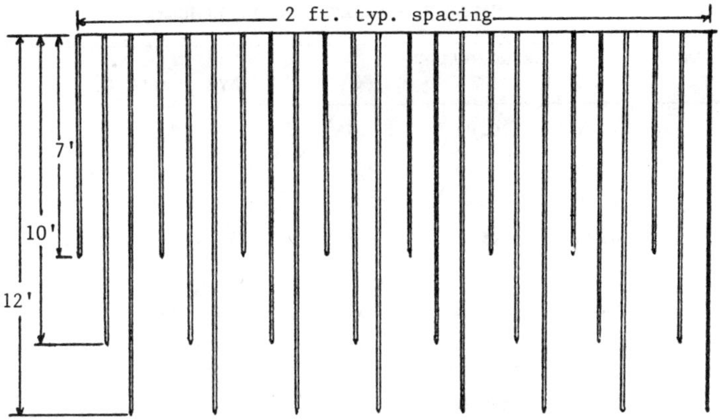

Fig. 5 - Cross Section of Soil Mass Indicating Staggered Depths of Injection

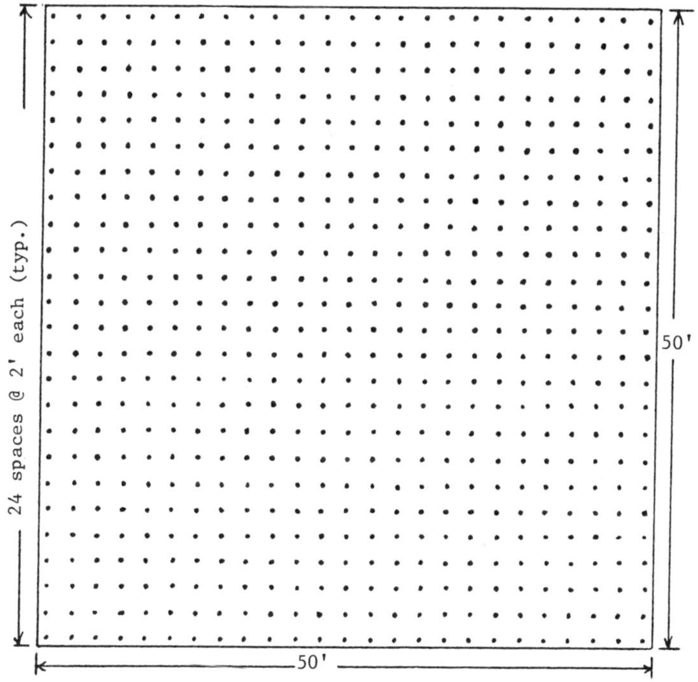

Fig. 6 - Spacing of Injection Points after 4 Injection Passes

TABLE I - INJECTED TEST RESULTS

DEPTH FT.	BORING NUMBERS 1	2	3	AVG. SWELL S_i%	S_i% CUM
1	1.7%	0.8%	0.4%	1.0% x 12" = 0.12	0.12"
2	1.5%	0.8%	0.4%	0.9% x 12" = 0.11	0.23"
3	1.2%	0.8%	0.9%	1.0% x 12" = 0.12	0.35"
4	0.8%	0.7%	1.8%	1.1% x 12" = 0.13	0.48"
5	0.6%	0.7%	1.8%	1.2% x 12" = 0.14	0.62"
6	0.6%	0.7%	0.1%	0.9% x 12" = 0.11	0.73"
7	0.6%	0.7%	0.1%	0.9% x 12" = 0.11	0.84"
		INTERFACE			
8	2.3%	2.9%	3.3%	2.8% x 12" = 0.34	1.18"
9	3.2%	2.4%	2.9%	2.8% x 12" = 0.34	1.52"
10	3.8%	2.4%	2.7%	3.0% x 12" = 0.36	1.88"
11	3.6%	2.3%	2.8%	2.9% x 12" = 0.35	2.23"
12	3.1%	2.3%	3.2%	2.9% x 12" = 0.35	2.58"

TABLE II - CONTROL TEST RESULTS

DEPTH FT.	BORING NUMBERS 4	5	AVG. SWELL S_i%	S_i% CUM
1	0.4%	1.0%	0.7% x 12" = 0.08	0.08"
2	0.4%	2.0%	1.2% x 12" = 0.14	0.22"
3	1.2%	4.0%	2.6% x 12" = 0.31	0.53"
4	1.5%	6.0%	3.8% x 12" = 0.46	0.99"
5	2.3%	7.2%	4.8% x 12" = 0.58	1.57"
6	2.7%	7.2%	5.0% x 12" = 0.60	2.17"
7	2.7%	7.2%	5.0% x 12" = 0.60	2.77"
	INTERFACE			
8	2.7%	3.8%	3.3% x 12" = 0.40	3.17"
9	2.6%	3.3%	3.0% x 12" = 0.36	3.53"
10	2.5%	2.4%	2.5% x 12" = 0.30	3.83"
11	2.5%	1.8%	2.2% x 12" = 0.26	4.09"
12	2.6%	2.4%	2.5% x 12" = 0.30	4.39"

Utilizing the data from post-injection swell tests along with the layer seasonal factors, the TPS swell in inches was next calculated for each 12-in layer of the test pad. Table III shows the complete calculations of Total Probable Swell (TPS) using the new methods based on the theories of Davis and Tucker. The data indicates a TPS of 0.763 inches in the top 12-ft of the treated expansive clay building foundation soil. The design criteria of 1.0-in swell or less was satisfied.

TABLE III POST INJECTED SWELL CALCULATIONS

DEPTH FT.	SEASONAL FACTOR F*	POTENTIAL SWELL S	SEASONAL SWELL S	CUMULATIVE SWELL S TOTAL
1	1.00	0.12	0.12	0.12
2	0.82	0.11	0.0902	0.2102
3	0.69	0.12	0.0828	0.293
4	0.59	0.13	0.0767	0.3697
5	0.50	0.14	0.0700	0.4397
6	0.40	0.11	0.044	0.4837
7	0.31	0.11	0.0341	0.5178
8	0.27	0.34	0.0918	0.6096
9	0.18	0.34	0.0612	0.6708
10	0.12	0.36	0.0432	0.7140
11	0.09	0.35	0.0315	0.7455
12	0.05	0.35	0.0175	0.763

*Davis-Tucker, 1973

SUMMARY AND CONCLUSIONS

The test pad final results based on post-injection swell test data showed that the swell of the injected soil was less then the swell in the control section, and the final moisture content of the injected soil was less than the final moisture content of the control soil. This is explained by the clay having its exchange capacity saturated with the calcium ion. When this occurs the calcium saturated clay will not swell beyond an interlayer spacing of approximately 10 angstroms, whereas, the same clay saturated with the sodium ion will swell, under no applied load or overburden, until its interlayer spacing is about 500 angstroms. The amount of swelling will depend, of course, on the applied load or overburden. When a clay is more than 15 percent saturated with the sodium ion, it acts largely like it is sodium-saturated. This is seen as positive evidence of the physicochemical changes brought about as a result of the multiple lime injections.

The conclusions from the field demonstration of the first STABIL-PAD, LSPI stabilization project conducted in Irving, Texas are as follows:

1. Pre-injection testing can be a viable part of the LSPI process.

2. A maximum injection spacing of 2 ft is required for effective expansive clay stabilization.

3. Injection depth must exceed the depth of active swelling.

4. The amount of lime slurry required for stabilization is several times that which was previously used.

5. Post-Injection Swell Test are valid for acceptance testing of injected sites.

6. An analytical method of predicting total probable swell (TPS) is required to understand LSPI results.

Building foundations constructed on expansive clay soils have long been a challenge to the engineers, architects and contractors. The LSPI industry was begun over 20 years ago as one answer for the treatment of these problem foundation soils. During the years since its inception a variety of small changes and inovations have been developed and added to the technology by those involved to increase the reliability and effectiveness of the method. In 1985 the state-of-the-art of lime slurry pressure injection was documented and published in NLA Bulletin 331, Boyton and Blacklock (1985). Also, in 1985 injection stabilization engineering research and development forces began work on a much improved new system of LSPI that would incorporate multiple injections, closer spacing, pre-injection testing, post-injection acceptance tests, rational analysis, engineering design and a contractor's guarantee. The results of this effort including the reporting of a full scale test pad have been presented.

The information presented in this paper has taken the reader from the beginning days of LSPI research, development and construction when it was the new promising method of expansive clay stabilization 25 years ago to the present. A new method of LSPI stabilization called STABIL-PAD has been discussed and laboratory and field data for a fullscale test pad using this new method in Irving, Texas have been presented. The post injection data from the test pad was analyzed for total probable swell and the method of analysis and results of the analysis are presented. The new Irving Mall extension buildings were completed in the Fall of 1987. Considerable cost savings were achieved as a result of using this new slab stabilization method.

REFERENCES

Blacklock, J. R., Improvement of Problem Track Subsoil by the Lime Slurry Pressure Injection Method, Final Report, Federal Railroad Administration, DOT-OS-40107, June, 1977.

Blacklock, J. R., "Lime Slurry Glaze Method of Soil Stabilization Testing," Proceedings - ASCE Specialty Conference on Grouting in Geotechnical Engineering, New Orleans, Louisiana, February, 1982.

Boynton, R. S., and Blacklock, J. R., Lime Slurry Pressure Injection Bulletin, National Lime Association, Arlington, Virginia, Bulletin 331, 1985.

Building Research Advisory Board, "Chemical Soil Stabilization," National Academy of Sciences, Washington, DC, 1969.

Das, B. M., Principles of Foundation Engineering, Brooks/Cole Engineering Division, Wadsworth, Inc. 1984.

Davis, R. C., and Tucker, R. L., "Soil Moisture and Temperature Variations Beneath a Slab Barrier on Expansive Clay," Construction Research Center, University of Texas at Arlington, report no TR-3-73, May, 1973.

Fenner, J. L., Hamberg D. J., and Nelson, J. D., "Building on Expansive Soils," Civil Engineering Department, Colorado State University, 1983.

Fohs, D. G., and Kinter, E. B., "Migration of Lime in Compacted Soil," Public Roads, 37, No. 1, June, 1972.

Haston, J. S., "Use of Lime Slurry Pressure Injection in Expansive Soil," Texas Section Fall Meeting Paper, ASCE, 1981.

Haynes, J. H., and Mason, R. C., "Subgrade Soil Treatment of the Apparel Mart, Dallas, Texas," Proceedings, Second International Research and Engineering Conference on Expansive Clay Soils, Texas A & M University, 1969.

Higgins, C. M., "High Pressure Lime Injection," Research Report No. 17, Interim Report No. 2, Research Project No. 63-7S, Louisiana Department of Highways, August, 1965.

Higgins, C. M., "Lime Treatment at Depth," Research Report No. 41, Final Report, Research Project No. 63-7S, Louisiana Department of Highways, June, 1969.

Ingles, O. G., & Neil, R. C., "Lime Grout Penetration and Associated Moisture Movements in Soil," Research Paper No. 138, Division of Applied Geo-mechanics, C.S.I.R.O., Australia, 1970.

Jones, D. E., "Expansive Soils and Housing Developments," Presented to the Workshop on Expansive Clays and Shales in Highway

Design, Denver, Colorado, December 1972.

Jones, D. E., and Holtz, Wesley, G., "Expansive Soils - The Hidden Disaster," Civil Engineering - ASCE, August, 1973.

Jones, D. E., and Jones, K. A., "Treating Expansive Soils," Civil Engineering - ASCE, August, 1987.

Lundy, H. L. Jr., and Greenfield, B. J., "Evaluation of Deep In-Situ Soil Stabilization by High-Pressure Lime-Slurry Injection," Highway Research Record No. 235, Highway Research Board, 1968.

McKeen, G. R., and Hamberg, D. J., "Characterization of Expansive Soils," Transportation Research Record, No. 790, 1981.

McKeen, G. R., "Design of Airport Pavements for Expansive Soils," Department of Transportation report no. FFA/RD-87/25, January, 1981.

Petry, T. M., Armstrong, J. C., and Chang, T., "Short Term Active Soil Property Changes Caused by Injection of Lime and Fly Ash," Proc., Transportation Research Board, Washington, DC, 1982.

Thompson, M. R., "Influence of Soil Properties on Lime-Soil Reactions," Public Works, August, 1965.

Thompson, M. R., "Field Evaluation of Pressure Injection Lime Treatment for Strengthening Subgrade Soils," Proceedings of Roadbed Stabilization Lime Injection Conference, August, 1975, FRA-OR&D-76-137, November, 1975.

Thompson, M. R., and Robnett, Q. L., "Pressure Injection Lime Treatment of Swelling Soils," Presented at Transportation Research Board 54th Annual Meeting, Washington, DC, January, 1975.

USE OF IN SITU TESTS TO PREDICT UPLIFT PERFORMANCE OF MULTIHELIX ANCHORS

by

Alan J. Lutenegger[1], M. ASCE, Bradford L. Smith[2] and Mohammed G. Kabir[2], A.M. ASCE

ABSTRACT

To investigate the applicability of in situ tests for design of multihelix anchors, tests were conducted at three sites where full-scale uplift tests had previously been conducted and reported in the literature. The three sites consisted of dense fine beach sand, sensitive marine clay and lacustrine varved silt. In situ tests conducted at the sites included: electric cone penetrometer, piezocone penetrometer, Marchetti Dilatometer, Borehole Shear Test, Pencel Pressuremeter, Menard three-cell and monocell pressuremeters and vane shear. Anchor tests had been performed with relative H/D ratios ranging from 4 to 12 to investigate effects of relative embedment and included both short-term and long-term tests. A comparison is made between the ultimate vertical uplift capacity obtained from load tests and predictions from the results obtained from each of the in situ tests.

INTRODUCTION

Multihelix anchors provide an economical alternative for resisting uplift or pullout loads for a variety of applications, and may be particularly desirable to the electric power industry for foundations of tower structures (Weikart and Clemence, 1987; Rodgers, 1987). Quite often, it is impractical to conduct traditional site investigations at each transmission tower location, especially when transmission lines traverse large areas and difficult terrain and most likely include different geologic conditions.

In the past decade, the rapid growth of in situ testing has produced an increased awareness to practicing geotechnical engineers of the large number of test devices available. However, quite often much confusion is associated with understanding limitations and applicability of different tests for specific design requirements and

[1]Assoc. Prof. of CEE, Clarkson University, Potsdam, N.Y. 13676

[2]Res. Asst., Dept. of CEE, Clarkson University, Potsdam, N.Y. 13676

soil types. For mostly economic reasons, it appears that there is a high degree of desirability on the part of many design engineers to implement some form of in situ testing as a part of routine site investigation programs. This probably is related to the advantage of having a larger data base in a shorter period of time. Thus, the need for unnecessary overconservatism, which results from having limited design data, is reduced. The application of in situ tests to the design of anchors may be appealing considering that a relatively small volume of soil in a relatively isolated zone is involved in the failure.

In this paper, a comparison is made between the results of full scale uplift load tests conducted on multihelix anchors and predictions of ultimate uplift load made from the results of a number of in situ tests.

FIELD TEST SITES

The work reported herein was conducted at three sites where full-scale multihelix anchor pullout tests had previously been conducted and reported (Mooney et al. (1985), Mitsch and Clemence (1985)). A summary of the site profiles is shown in Fig. 1 and a brief description of soils encountered at each test site follows.

Bryant Core Sand (BCS)

As described by Mitsch and Clemence (1985) this site is located approximately 80 Km north of Syracuse, N.Y. and consists of a large deposit of sand which represents a remnant beach deposit of Pleistocene age. Throughout the site in about the upper meter, drilling and in situ testing was difficult because of cobble size erratics which appeared to be randomly distributed. Below this depth and extending to a depth of about 5 m the sand appeared to be medium dense to dense and extremely uniform in grain size. No water table was found within the zone where field tests were conducted.

St. Lawrence Seaway (SLS)

This site was described by Mooney et al. (1985) as a medium stiff marine clay with numerous sand lenses extending to a depth of about 5.5 m. The site is located about 16 Km northeast of Massena, N.Y. and is about 1 Km north of the Snell Lock on the St. Lawrence Seaway. The site materials represent the upper portion of desiccated Leda Clay in this area. In the upper meter, the material is interbedded with sand lenses, which is typical for the general area. Below this depth, the texture and plasticity becomes more uniform as shown in Fig. 1. A water level at a depth of about 1 m was present at the site during testing.

Fulton (FLT)

This site was also described by Mooney et al. (1985) and is located about 0.8 Km northeast of Fulton, N.Y. The soils consist of about 4.5 m of varved lacustrine silt and clay with abundant interbedded sand seams, all of presumed glacial origin, underlain by dense glacial

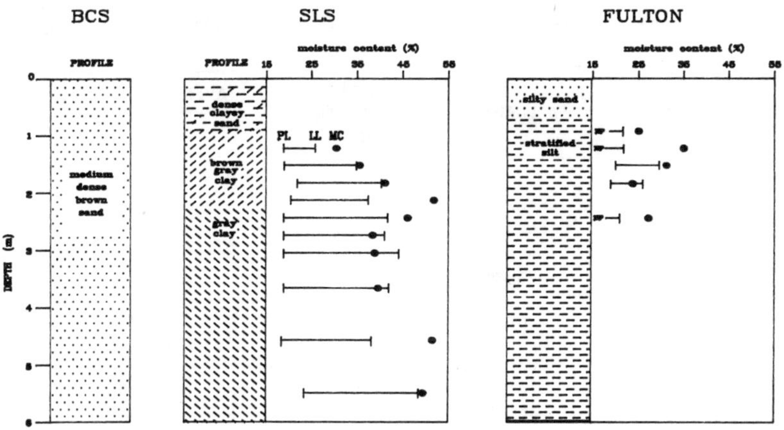

Fig. 1. Site Characteristics

till. This site probably represents the most variable soil conditions of the three sites investigated. Groundwater observations at the site indicated a water level at about 1.4 m.

MULTIHELIX ANCHOR TESTS

The results of full-scale multihelix anchor uplift tests conducted at the three sites have been described in detail by Mooney et al. (1985) for the SLS and FLT sites and by Mitsch and Clemence (1985) for the BCS site. Both short-term and long-term tests were conducted using triple-helix anchors of geometry shown in Fig. 2. A summary of the test results which includes tests conducted at all three sites is shown in Table 1. Only tests conducted in vertical uplift have been included for the current study.

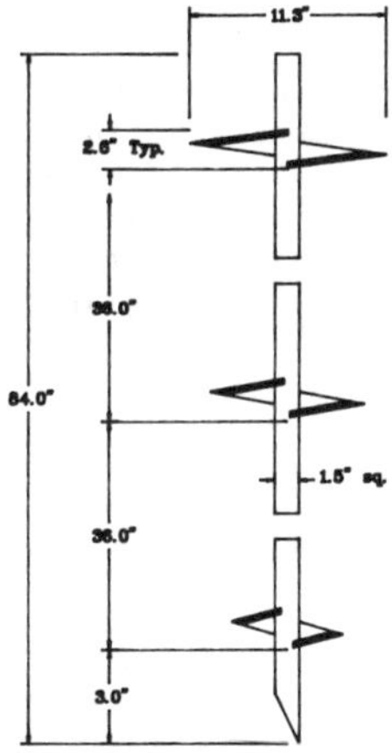

Fig. 2. Typical Anchor Geometry (Mitsch & Clemence, 1985)

TABLE 1. Summary of Anchor Tests and Reported Uplift Load

SITE	LOADING	RELATIVE DEPTH H_1/D_1*	AVERAGE ULTIMATE UPLIFT LOAD (pounds)**
BCS	Short Term	4	36,750
	Short Term	8	49,250
	Long Term	4	32,000
	Long Term	8	45,000
SLS	Short Term	4	12,000
	Short Term	8	11,350
	Short Term	10	10,850
	Short Term	12	10,800
	Long Term	4	14,000
	Long Term	8	14,000
	Long Term	10	14,000
FLT	Short Term	4	23,000
	Short Term	8	24,000
	Long Term	4	17,000
	Long Term	8	24,500

* H_1 = depth to top helix
  D_1 = diameter of top helix

** 1 lb = 4.45N

IN SITU TESTING

At each of the sites investigated, a series of in situ tests was conducted to investigate the relative applicability of each of the tests for each soil type in different loading modes. Because of differences in soil type, groundwater conditions, site accessibility, equipment availability and scheduling, it was not possible or practical to conduct the same number of tests at the same depths at all three sites. Table 2 shows a summary of tests conducted at each site. A detailed plan showing the frequency of each in-situ test at each of the sites is shown in Fig. 3. A brief description of the procedures and interpretation of the results will be presented for each of the tests.

Cone Penetration Test (CPT)

Continuous site profiling using an electric cone penetrometer to measure tip resistance, q_c, was conducted using a conventional 35.7 mm diameter 60° apex electric cone. The procedure described by ASTM D3441 (ASTM, 1987) was followed. For the SLS site, a piezocone with the piezo element located 5.7 mm behind the cone base was also used to measure penetration pore pressures. CPT profiles for all three sites are shown in Fig. 4.

TABLE 2. Summary of In Situ Tests Conducted

SITE	TEST					
	CPT[1]	DMT[2]	PMT[3]	BST[4]	VST[5]	PCL[6]
BCS	X	X	X	X		
SLS	X	X	X*	X	X	X
FLT	X	X	X	X		

*Menard three-cell PMT
[1]Quasistatic Cone Penetration Test
[2]Flat Dilatometer Test
[3]Prebored Monocell Pressuremeter Test
[4]Borehole Shear Test
[5]Vane Shear Test
[6]Pencel Full Displacement Pressuremeter Test

DEPTH (FT)*	BCS					SLS						FLT			
	CPT	DMT	PMT	BST	PCL	CPT	DMT	PMT	BST	VST	PCL	CPT	DMT	PMT	BST
1						X	X						X		
2	X	X				X	X	X				X	X		
3	X	X				X	X		X			X	X	X	X
4	X	X	X	X		X	X	X	X	X	X	X	X	X	X
5	X	X				X	X	X	X	X		X	X	X	X
6	X	X	X	X	X	X	X	X	X	X	X	X	X	X	X
7	X					X	X	X	X	X		X	X	X	
8	X		X	X	X	X	X	X		X	X	X	X		X
9	X					X	X	X		X		X	X		
10	X		X	X		X	X	X		X	X	X	X		
11	X					X	X			X		X	X		
12	X		X	X		X	X			X	X	X	X		
13	X					X	X			X					
14	X		X	X		X	X			X					
15						X	X			X					
16						X	X			X	X				

*1 ft = .304 m

Fig. 3. In Situ Testing Plan for Each Site

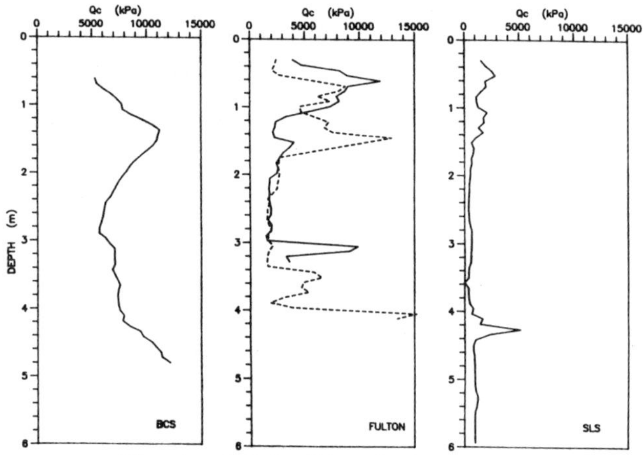

Fig. 4. CPT Profiles for Each Site

Estimates of undrained shear strength in the cohesive materials were made from the cone tip resistance using the empirical equation:

$$s_u = \frac{q_c - \sigma_{vo}}{N_k} \qquad [1]$$

where σ_{vo} = total vertical stress at the cone tip

N_k = an empirical factor

The value of N_k was taken as 16 and 20 for SLS and FLT, respectively based on average soil plasticity using the chart presented by Aas, et al. (1986).

Undrained strength estimates from piezocone pore pressure measurements were taken from:

$$s_u = \frac{\Delta u}{N} \qquad [2]$$

where Δu = increase in pore pressure from cone pentration, i.e. $u_{excess} - u_{in\ situ}$

N = an empirical factor

A value of N = 9.5 was used following the recent recommendation of Greig et al. (1987).

Estimates of effective friction angle from cone tip resistance measurements in sands were made using the charts recommended by Robertson and Campanella (1983). An estimate of ϕ' was made for each 0.3 m depth.

Dilatometer Test (DMT)

Marchetti Flat Dilatometer tests were conducted at SLS and FLT for the full profile and in the upper zone at BCS using the procedures recommended by Schmertmann (1986). Tests were conducted at 0.3 m intervals beginning at a depth of 0.3 m. Values of the DMT material index, I_D, were used to interpret site conditions and guide in selecting appropriate data reduction procedures. Estimates of undrained shear strength and coefficient of earth pressure at rest K_o, in cohesive soils were made using the empirical correlations presented by Marchetti (1980). Estimates of the effective friction angle and K_o in sands were made using the procedure described by Schmertmann (1982).

Pressuremeter Test (PMT)

Prebored and full-displacement pressuremeter tests were conducted at each of the sites. Prebored pressuremeter tests were performed using a 76 mm diameter monocell pressuremeter with a length/diameter ratio of 8, except at SLS, where a conventional three-cell Menard PMT was used. Testing procedures recommended by Winter (1982) were followed for all prebored PMT's. All tests were conducted in boreholes drilled using a 76 mm diameter closed-bucket hand auger.

The Pencel probe described by Briaud and Shields (1979) was used for full-displacement PMT tests. A stress controlled pressure console/volumeter assembly was used so that the test could be conducted in a manner similar to the prebored PMT. The probe was advanced into the ground under quasi-static penetration without a prebored hole. Tests were initiated immediately after penetration stopped at the desired test depth.

Estimates of undrained shear strength in clays were made from both tests using the interpreted limit pressure, P_L, and the conventional empirical equation:

$$s_u = \frac{P_L - P_o}{N_p} \qquad [3]$$

The magnitude of P_o was obtained from interpreting the creep volume-pressure curve from prebored tests. For the Pencel, the value of P_o was calculated using K_o values obtained from the DMT and measured water levels. The value of N_p was taken as 5.5 after a careful review of the literature (Lutenegger, 1987b).

Estimates of the effective friction angle in sands were made from prebored PMT results using the procedure presented by Hughes et al. (1977). K_o values were estimated from prebored PMT results using:

$$K_o = \frac{P_o - u_o}{\sigma'_{vo}} \qquad [4]$$

Borehole Shear Test (BST)

The procedure described by Lutenegger (1987a) was used to conduct BSTs. Multistage tests were conducted in 76 mm diameter hand augered boreholes. A consolidation time of 5 min. was used for all tests. Least squares linear regressions were performed to determine friction angle and cohesion.

Vane Shear Test (VST)

A Nilcon self-recording vane apparatus was used to conduct field vane tests at SLS. The procedure recommended by ASTM D2573 (ASTM, 1987) was followed.

DESIGN ANALYSIS

While a number of studies have been conducted on the ultimate uplift capacity of anchor plates in soils (e.g., Meyerhof and Adams, 1968) it appears that limited research has been performed on multihelix anchors. A design procedure for multihelix anchors was proposed by Mitsch and Clemence (1985) for sands and by Mooney et al. (1985) for anchors in silt and clay. The writers used these procedures to evaluate the ultimate uplift capacity. Discussion of specific design details are referred to in these publications and will only be briefly reviewed herein.

In sands, Mitsch and Clemence proposed that the failure for shallow and deep anchor installations would appear as shown in Fig. 5. The ultimate anchor capaicty was given as:

shallow ($H_1/D_1 \leq 4$)

$$Q_u = \pi\gamma' K_u \tan\phi_1 \cos^2(\phi_1/2)\left\{\frac{D_1H_1^2}{2} + \frac{H_1^3\tan(\phi_1/2)}{3}\right\} + \frac{\pi}{2}D_a\gamma'(H_3^2-H_1^2)K_u\tan\phi_2 + W_s \qquad [5]$$

deep ($H_1/D_1 > 4$)

$$Q_u = \gamma' H_1 A_1 N_{qu} + \frac{\pi}{2} D_a\gamma'(H_3^2-H_1^2)K_u \tan\phi_2' + P_s H_1\left(\frac{\gamma' H_1}{2}\right)K_u \tan\phi'_1 \qquad [6]$$

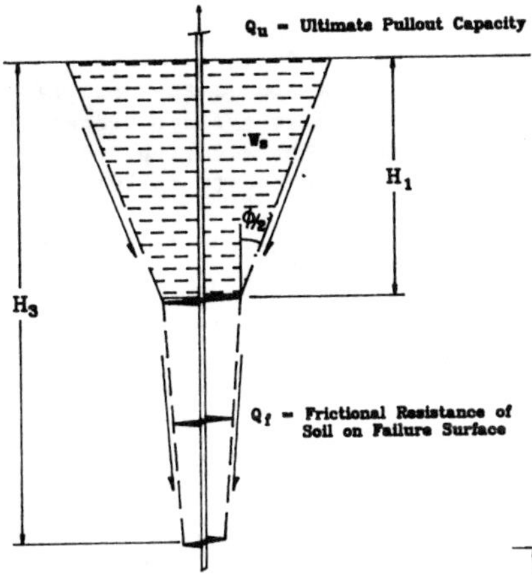

Fig. 5a. Proposed Failure Mode for Shallow Multihelix Anchors in Sand (Mitsch & Clemence, 1985)

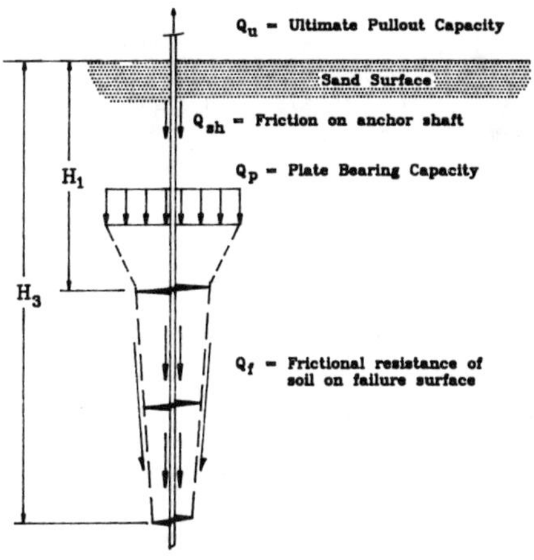

Fig. 5b. Proposed Failure Mode for Deep Multihelix Anchors in Sand (Mitsch & Clemence, 1985)

where Q_u = ultimate anchor uplift capacity

γ' = effective unit weight of soil

K_u = coefficient of lateral earth pressure in uplift for sands

ϕ_1 = friction angle of soil above H_1

ϕ_2 = friction angle from H_1 to H_3

A_1 = area of top helix

N_{qu} = uplift capacity factors for sands

H_1 = depth to top helix

H_3 = depth to bottom helix

D_a = average helix diameter

D_1 = diameter of top helix

P_s = perimeter of anchor shaft

W_s = weight of soil above the top helix

Equations (5) and (6) are slightly different than presented by Mitsch and Clemence (1985) in that they allow for different values of friction angle above and below the top helix. As shown in Fig. 5, the uplift failure of shallow anchors ($H_1/D_1 \leq 4$) assumes a wedge breakout above the top helix. In the case of a deep anchor installation ($H_1/D_1 > 4$) the top helix is assumed to develop a bearing capacity failure.

In clays and silts, a similar design procedure was proposed by Mooney et al. (1985). The proposed failure mode is shown in Fig. 6. The ultimate uplift capacity in undrained (ϕ=0) loading for clays is given as:

$$Q_u = A_1 s_u N_u + \pi D_a s_u (H_3 - H_1) + P_s H_1 c_a \quad [7]$$

In the case of silts, it was assumed that drainage may occur during loading, and therefore the uplift behavior would be more like that of cohesionless soils. Therefore, a contribution from the frictional resistance of soil is included. The ultimate uplift is given as:

$$Q_u = \gamma' H_1 A_1 N_{qu} + A_1 c N_u + \frac{\pi}{2} D_a \gamma' (H_3^2 - H_1^2) K_u \tan\phi' + \pi D_a c (H_3 - H_1) + P_s H_1 \frac{\gamma'}{2} H_1 K_u \tan\phi' + P_s H_1 c_a \quad [8]$$

where Q_u = ultimate anchor uplift capacity
γ' = effective unit weight of soil
H_1 = depth to top helix
H_3 = depth to bottom helix
N_{qu} = uplift capacity factor for cohesionless soils
N_u = uplift bearing capacity factor for cohesive soils
D_a = average helix diameter
c = cohesion
K_u = coefficient of lateral earth pressure in uplift cohensionless soils)
P_s = perimeter of anchor shaft
c_a = adhesion on anchor shaft
A_1 = area of top helix

The value of the uplift bearing capacity factors N_{qu} and N_u are a function of the relative embeddment, i.e. H_1/D_1, and of soil strength. Summaries of appropriate values for these factors are presented by Mooney et al. (1985).

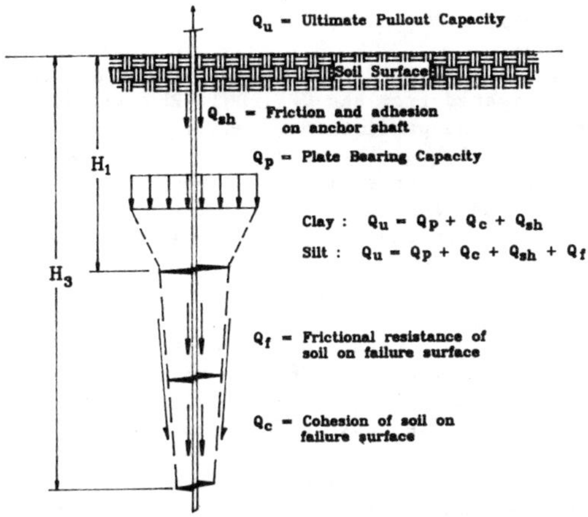

Fig. 6. Proposed Failure Mode for Anchors in Clay and Silt (Mooney et al., 1985)

RESULTS

Considering the variation in the types and number of in situ tests conducted at each site, there are several combinations which could be used to predict the uplift capacity. For uplift capacity in sands, the pertinent soil parameters are friction angle and uplift coefficient of lateral earth pressure; in clays the required parameters are undrained shear strength and adhesion for short term loading; in silts, friction angle, cohesion and adhesion are necessary (Mooney et al., 1985; Mitsch and Clemence, 1985). Long term loading in clays may allow some drainage to take place and thus there may be a component of friction which contributes to the ultimate capacity. In this case, the design methodology would change and the use of Eq. (8) may be more appropriate. A summary indicating which test results were used for estimating uplift at each site is presented in Table 3.

TABLE 3. Application of In Situ Tests to Loading and Site Conditions

Site	Behavior	CPT	DMT	PMT	BST	VST	PCL
BCS	drained	X	X	X	X		
SLS	Short Term (undrained)	X	X	X		X	X
	Long Term (drained)				X		
FLT	Short Term (undrained)	X	X	X			
	Long Term (drained)				X		

Anchor Behavior in Sand

Calculations of ultimate uplift load in the sand at BCS were made using ϕ' values estimated from the CPT, DMT, PMT and BST. In order to assess the accuracy of the predictions, individual values of ϕ' were used in Eqs. 5 and 6 and the value of K_u was backcalculated using measured uplift loads for both shallow and deep anchors. The average of the short-term and long-term field tests was used in these calculations. A summary of the results is presented in Table 4.

Results from both the PMT and BST gave values of K_u considerably higher than suggested by Mitsch and Clemence (1985) primarily since the values of ϕ' were much lower than obtained from the CPT and DMT. Values of ϕ' obtained from the CPT and DMT were very close to each other with a mean of about 44°. Friction angles measured with the PMT gave a mean of 29° while the BST values averaged 32°. Based on standard penetration tests, a value of ϕ' = 35-40° was given by Mitsch and Clemence (1985) for this site. Since both the BST and PMT were performed in prebored hand-auger boreholes, it is reasonable to expect some loosening of the sand and thus a reduction in the measured values of ϕ'.

For a given value of H_1/D_1 the bearing capacity factor N_{qu} is a function of the effective friction angle, therefore a reduction in ϕ' also leads to a reduction in N_{qu}. The result is that a larger value of K_u is required to make predictions match observations. Backcalculated values of K_u assume undisturbed soil conditions such that $\phi_{insitu} = \phi_{uplift}$. Anchor installation effects are taken into account in the recommended K_u values.

The prediction of deep anchor capacity based on the CPT gave an unrealistically low backcalculated value of K_u. It is apparent that adjustment to either the friction angles or design approach is required to match measured uplift.

TABLE 4. Comparison between Backcalculated and Recommended Values of K_u in Sand (BCS).

TEST	H_1/D_1	Backcalculated K_u	Recommended K_u
CPT	4	2.26	2.0
	8	0.29	3.2
DMT	4	2.15	2.0
BST	4	3.79	0.9
	8	2.84	1.1
PMT	4	3.61	0.9
	8	2.51	1.1

Anchor Behavior in Clay

A comparison between the measured and predicted ultimate uplift capacity of short-term loaded anchors at SLS is shown in Fig. 7. Predicted values using interpreted undrained shear strength and Eq. 7 are generally much higher than measured capacities, except for the DMT and CPT pore pressure measurement which are both known to underpredict s_u in stiff and overconsolidated clays. The poor agreement between measured and predicted capacities cannot be explained by errors in interpretation of s_u from some of the tests which involve empirical correlation factors (e.g., Eqs. 1, 2 and 3) since the required factors would be well outside the established range of realistic values. The fact that all of the other tests overpredict the measured capacity suggests that it may be appropriate to modify the design approach. This may be especially true when designing anchors in sensitive marine clays.

The rotary action during installation of the anchors may cause substantial remolding of the soil in the zone between the top and bottom helix. Additionally, since the anchor shaft is square, the rotary action would produce a void space along the shaft above the top helix. It therefore seems appropriate to eliminate shaft adhesion and use remolded shear strength below the top helix.

If one considers these two possibilities the ultimate undrained capacity results only from the undisturbed strength above the top helix and remolded strength below the top helix. A comparison between the measured and predicted capacities using this modified procedure and average sensitivity from the field vane is shown in Fig. 8. As can be seen, much closer predictions are obtained from all tests.

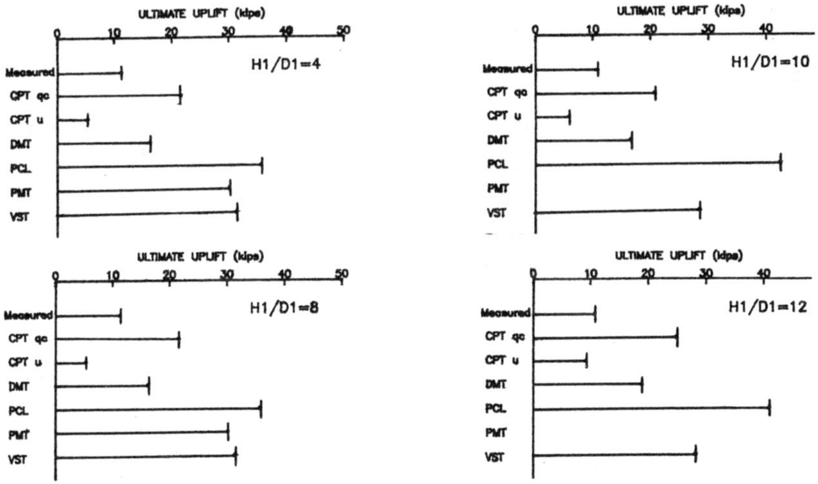

Fig. 7. Comparison Between Predicted and Measured Short-Term Loading Uplift Capacity in Clay

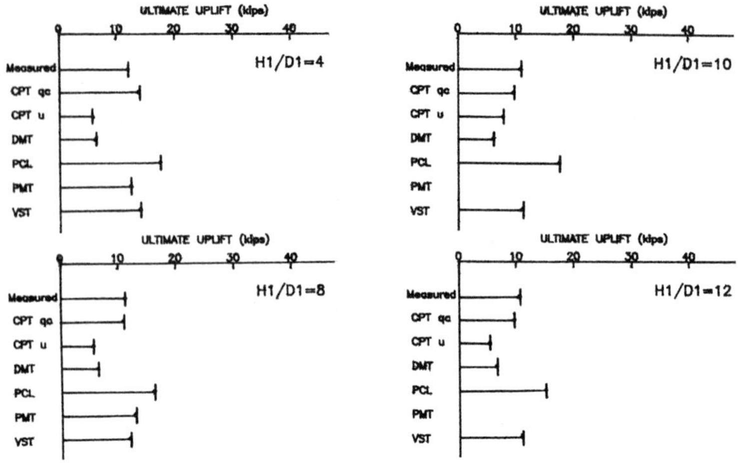

Fig. 8. Comparison Between Predicted and Measured Uplift Capacity in Clay (Modified Technique)

Anchor Behavior in Silt

Ultimate uplift capacities in silt were calculated for short-term loading behavior assuming undrained shear (ϕ=0) conditions. Predictions were made using Eq. 7, in which case the undrained shear strength is the only soil parameter required. Predictions were made using values obtained from the CPT, DMT and PMT for H_1/D_1 = 4 and from the CPT and DMT for H_1/D_1 = 8. These results are shown in Fig. 9 and generally indicated much better agreement with the measured capacities than had been found for the clay site. This may suggest that minimal disturbance has occurred during anchor installation and the adjustment for remolding effect previously applied to the sensitive clay site may be justified.

Long-term loading capacity was predicted using Eq. 8 and the results from the BST. A prediction was only made for H_1/D_1= 4, since deeper BST tests were not conducted. As with the predictions made in sand, the values of ϕ and c were incorporated into Eq. 8 and then the value of K_u was backcalculated for a comparison with suggested parameters. A value of K_u = 3.74 was obtained in this case, which is considerably higher than the recommended value of K_u = 0.9. While the ϕ' values are reasonable for silt (ϕ' = 32°) the BST is known to underpredict cohesion in most cases, which may account for the discrepancy noted.

It is interesting to note that if Eq. 8 and the BST data obtained at SLS are applied to the long-term shallow anchor test H_1/D_1 =4 at that site, a backcalculated value of K_u = 4.66 is obtained, which again is much higher than generally recommended. Thus, it appears that the BST prediction of long-term loading behavior of shallow anchors in cohesive soils requires adjustment for cohesion obtained in the field. Tentatively, an increase of about 150% appears appropriate to give realistic values of K_u. This analysis assumes fully drained conditions, which may not have been the case.

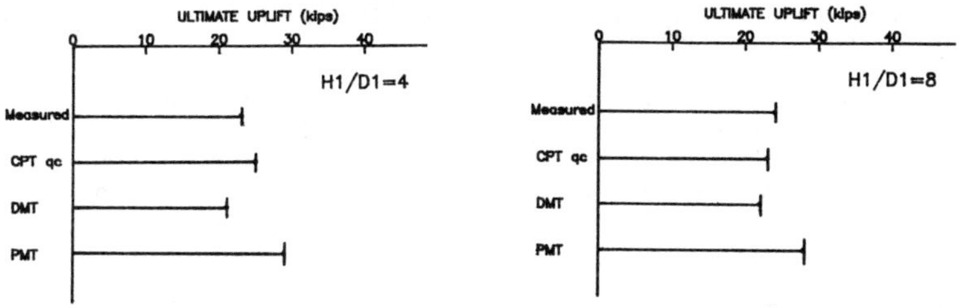

Fig. 9. Comparison Between Predicted and Measured Short-Term Loading Uplift Capacity in Silt

CONCLUSIONS

In situ tests conducted at three test sites were used to predict ultimate uplift capacity of multihelix anchors using published design procedures. The in situ tests provided site specific conventional design parameters for individual predictions for both shallow and deep anchors under short-term and long-term loading. Based on these analyses, the following tentative conclusions are presented regarding the use of in situ tests for determining uplift capacity of anchors:

Current proposed design methods for predicting ultimate uplift of multihelix anchors in sand, silt and clay provided reasonable predictions in most cases for <u>shallow</u> anchors with the following qualifications:

<u>sand</u>

a) The CPT and DMT provided reasonable results, generally within about 10%.

b) The BST and prebored PMT tended to underpredict capacity which may partially derive from loosening effects caused by predrilling.

<u>clay</u>

a) The short-term loading behavior of anchors in clay was closely predicted using undrained strength from the DMT and CPT pore pressure. Other tests, i.e., PMT, PCL, VST and CPT q_c overestimated uplift by a large degree using proposed design techniques.

b) Using a modified approach, which incorporates remolded strength parameters in the zone between the top and bottom helixes and eliminates shaft adhesion to account for anchor installation, produced very close agreement by PMT, PCL, VST and CPT q_c. This modification may only apply to sensitive clays and should be investigated further.

c) Long-term behavior of anchors in clay predicted using the BST assumed drained conditions and was on the low side of measured values. This may also be related to the characteristic low cohesion values measured by the BST.

silt

a) Short-term loading uplift capacity using CPT q_c, PMT, and DMT all provided reasonably accurate estimates.

b) Long-term behavior using the BST were similar to those obtained in clay, probably for similar reasons.

Deep anchor capacities predicted from in situ tests were close in the case of undrained analyses (modified clay method and silt) however for drained loading, the predictions were generally too high, especially in the case of sands.

The variation in accuracy obtained between measured and predicted performance for different soil types suggests that some modification to exisitng design procedures or test interpretations would be appropriate. While this paper has only considered proposed design techniques, other design philosophies for anchor behavior (e.g., Kulhawy, 1985) should be investigated for accuracy. Additionally, the development of specific design procedures relating to specific in situ tests may be deisrable. Such developments are currently underway.

REFERENCES

American Society for Testing and Materials, Annual Book of Standards, Vol. 04.08, 1987.

Briaud, J.-L. and Shields, D.H., "A Special Pressuremeter and Pressuremeter Test for Pavement Evaluation and Design, Geotechnical Testing Journal, ASTM, Vol. 2, 1979, pp. 143-151.

Hughes, J.M.O., Wroth, C.P. and Wintle, D., "Pressuremeter Tests in Sands," Geotechnique, Vol. 27, 1977, pp. 455-477.

Kulhawy, F.H., "Uplift Behavior of Shallow Soil Anchors - An Overview," Uplift Behavior of Anchor Foundations in Soil, ASCE, 1985, pp. 1-25.

Lutenegger, A.J., "Suggested Method for Performing the Borehole Shear Test," Geotechnical Testing Journal, ASTM, Vol. 10, 1987a, pp. 19-25.

Lutenegger, A.J., "Use of In Situ Tests to Determine Design Parameters for Drilled Shaft Foundations," Dept. of CEE Rept. No. 87-4, Clarkson Univ., 1987b.

Marchetti, S., "In Situ Tests by Flat Dilatometer," Jour. of the Geotechnical Engineering Division, ASCE, Vol. 106, 1980, pp. 299-321.

Mitsch, M.P. and Clemence, S.P., "The Uplift Capacity of Helix Anchors in Sand," Uplift Behavior of Anchor Foundations in Soil, ASCE, 1985, pp. 26-47.

Mooney, J.S., Adamczak, S. Jr., and Clemence, S.P., "Uplift Capacity of Helical Anchors in Clay and Silt," Uplift Behavior of Anchor Foundations in Soil, ASCE, 1985, pp. 48-72.

Robertson, P.K. and Campanella, R.G., "Interpretation of Cone Penetration Tests. Part I: Sand," Can. Geot. Jour., Vol. 20, 1983, pp. 718-733.

Rodgers, T.H., "High Capacity Multi-Helix Screw Anchors for Transmission Line Foundations," Foundations for Transmission Line Towers, ASCE, 1987, pp. 81-95.

Schmertmann, J.H., "A Method for Determining the Friction Angle in Sands from the Marchetti Dilatometer Test," Proc. 2nd European Symp. on Penetration Testing, Vol. 2, 1981, pp. 853-861.

Schmertmann, J.H., "Suggested Method for Performing the Flat Dilatometer Test," Geotechnical Testing Journal, ASTM, Vol. 9, 1986, pp. 93-101.

Weikart, A.M. and Clemence, S.P., "Helix Anchor Foundations - Two Case Histories," Foundations for Transmission Line Towers, ASCE, 1987, pp. 72-80.

Winter, E., "Suggested Practice for Pressuremeter Testing in Soils," Geotechnical Testing Journal, ASTM, Vol. 5, 1982, pp. 85-88.

MODEL TESTS FOR SHALLOW STRIP FOUNDATION ON GRANULAR SOIL

Braja M. Das[1] and Adel M. Hanna[2]

Abstract

Laboratory model test results for the ultimate bearing capacity of shallow strip foundation supported by a granular trench made in a weak saturated clay (ϕ=0 condition) have been presented. The ratio of the width of the granular trench to the width of the foundation has been varied from zero to two. Also the ratio of the height of the granular trench to the width of the foundation has been varied from zero to five. The model test results have been compared with the theory of Madhav and Vitkar (1978).

Introduction

Shallow foundations, when constructed over weak clay layers, may undergo large settlements. The load-bearing capacities of these foundations would be quite low. For that reason, several soil improvement techniques such as preloading, chemical stabilization, and use of stone columns under the foundation are now being utilized for improving the load-bearing capacity of the foundation and reducing the settlement at ultimate and allowable loads. The use and analysis of stone columns under shallow foundations have been published by several investigators (1,5,6,8). The construction of a stone column generally consists of waterjetting a vibrofloat into the soft clay layer to obtain a circular hole. The hole is then filled with an imported granular material. The granular material is compacted in the hole as the vibrator is withdrawn.

The concept of stone columns is schematically shown in Fig. 1. An isolated shallow circular foundation is being supported by the stone column. By means of field tests and small-scale laboratory model tests, Hughes, Withers, and

---

[1]Professor and Chairman, Department of Civil Engineering and Mechanics, Southern Illinois University at Carbondale, Carbondale, Illinois

[2]Associate Professor, Department of Civil Engineering, Concordia University, Montreal, Quebec, Canada

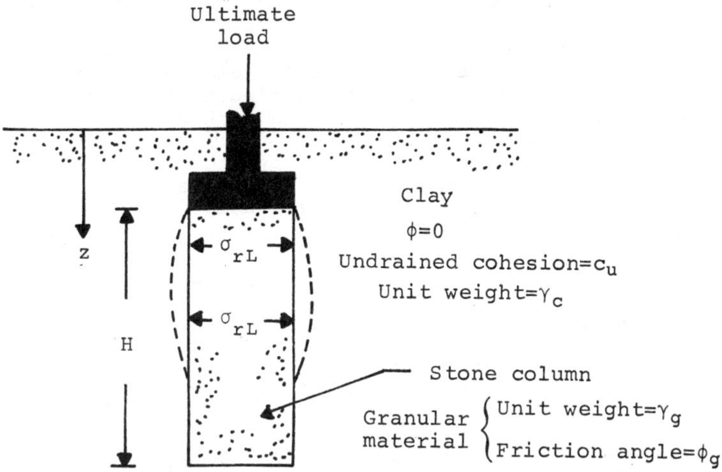

Fig. 1. Stone column supporting a circular foundation

Greenwood (6) have shown that, at ultimate load, the mode of failure of the stone column is by bulging as shown by the broken lines in Fig. 1. The ultimate bearing capacity of the foundation, thus, can be given as (6)

$$q_u = \sigma_{rL} \frac{1+\sin\phi_g}{1-\sin\phi_g} \tag{1}$$

in which q_u=ultimate bearing capacity of the foundation, ϕ_g= angle of friction of the granular material; and σ_{rL}=lowest value of the restraining passive pressure offered by the clay along the circumference of the stone column.

The magnitude of σ_{rL} has been given by Hughes, et al. (6) as

$$\sigma_{rL} = 4c_u + \gamma_c z \tag{2}$$

in which c_u=undrained shear strength of the saturated clay, γ_c=unit weight of clay, and z=depth below the ground surface.

The use of a granular trench in the construction of a shallow continuous foundation is a plane strain variation of the stone column approach discussed above. The present study relates to the laboratory determination of the ultimate bearing capacity of a shallow continuous foundation supported by a granular trench made in a weak clay.

Bearing Capacity Theory for Continuous Foundation on a Granular Trench

At the present time, very few studies relating to the ultimate bearing capacity of shallow continuous foundations supported by a granular trench made in a weak clay (either theoretical of experimental) have been published in the literature. Madhav and Vitkar (7), in order to obtain a theoretical relationship for such a case, assumed a general shear type of failure surface under the foundation as shown in Figs. 2a and 2b. In these figures, the width of the foundation is equal to B and the width of the granular trench is W. The failure zone consists of four parts. They are:

1. Rankine active zone--ABC;
2. Mixed transition zones--ACG and BCD;
3. Transition zones--AGI and BDE; and
4. Passive zones--AIJ and BEF.

CD and CG are arcs of a log spiral based on the friction angle of the trench material (ϕ_g). Similarly, DE and GI are arcs of a log spiral based on the friction angle of the clay (for this case, ϕ=0).

The theoretical ultimate bearing capacity, q_u, of the foundation was determined by using the upper bound limit analysis theorem (3,4). The ultimate bearing capacity relationship thus determined can be expressed as

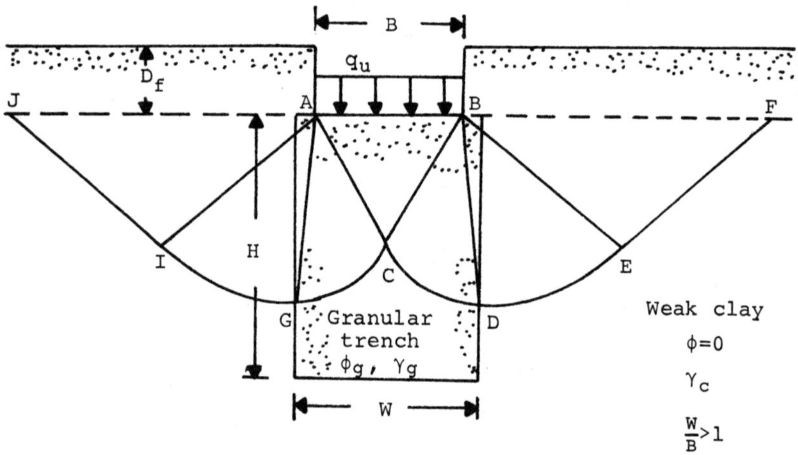

(a)

Fig. 2. Failure surface under a continuous foundation as assumed by Madhav and Vitkar (7): (a) for W/B>1, and (b) for W/B<1.

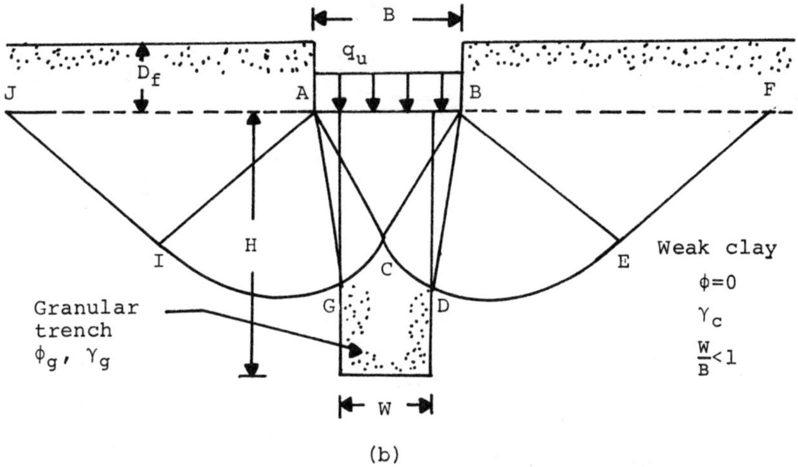

Fig. 2. Continued.

$$q_u = c_u N_c^* + qN_q^* + \frac{1}{2}\gamma_c BN_\gamma^* \quad (3)$$

in which $q=\gamma_c D_f$; D_f=depth of foundation measured from the ground surface; c_u=undrained shear strength of clay; γ_c=unit weight of clay; N_c^*, N_q^*, and N_γ^*=modified bearing capacity factors.

The variations of the modified bearing capacity factors with ϕ_g and W/B are given in Figs. 3, 4, and 5.

Laboratory Model Tests

In order to compare the theoretical findings of Madhav and Vitkar (7) with the actual performance, a number of small-scale model tests were conducted in the laboratory.

A large amount of a clayey soil was collected from the field. The soil was pulverized well in the laboratory. A wash sieve analysis was conducted on a representative specimen of this soil, and the grain-size distribution as determined in the laboratory is given in Fig. 6. The liquid and plastic limits of this clay were determined to be 31% and 18%, re-specitively. The granular material used for the trench was a poorly-graded sand, the grain-size distribution of which is also shown in Fig. 6. A predetermined amount of water was added to and thoroughly mixed with the clay. The moist

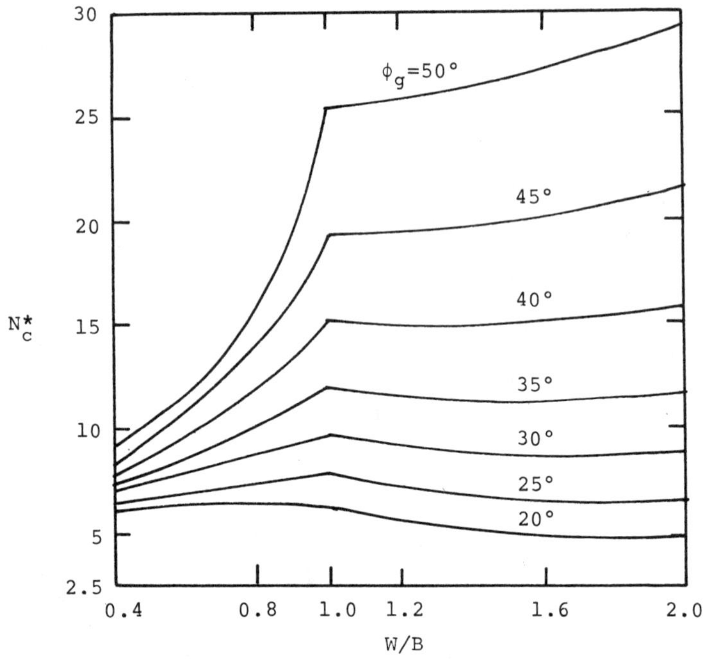

Fig. 3. Variation of N_c^* with ϕ_g and W/B based on the solution of Madhav and Vitkar

soil was then transferred to several plastic bags that were tightly sealed. These bags were stored in a moist curing room for about a week before use.

The model foundation was a steel plate with dimensions of 304.8 mm (length) × 101.6 mm (width) × 9.5 mm (thickness). The inside dimensions of the model test box were 914.4 mm (length) × 304.8 mm (width) × 914.4 mm (height). The sides of the box were heavily braced to avoid lateral yielding. The inside of the box was smoothed as much as possible with sandpaper to minimize friction during the tests.

For conducting the model tests only on sand or clay (without the granular trench), the soil was compacted in layers 50.8 mm to 76.2 mm thick in the model test box to the desired height. The compaction was done in sections by using a flat bottomed rammer. After compaction, the model foundation was centrally placed in the box on top of the compacted soil layer. The load on the model foundation was applied at a very slow rate to simulate static loading condition by using a hydraulic jack. The load and corresponding settlement

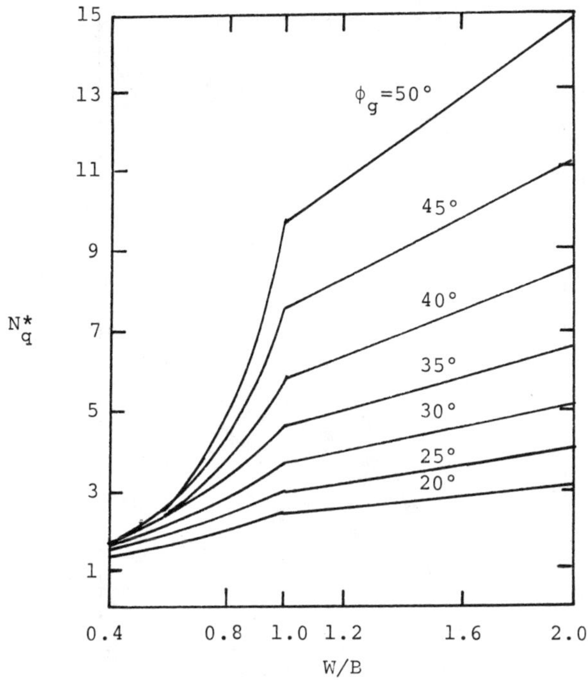

Fig. 4. Variation of N_q^* with ϕ_g and W/B based on the solution of Madhav and Vitkar

were recorded by a proving ring and dial gauge, respectively.

In order to conduct model tests in clay with a granular trench, the clay was first compacted in the test box in 50.8 mm layers to a desired height. For creating the granular trench, four wooden molds having outside dimensions of 304.8 mm (length) × W (width) × 508 mm (height) were used. The magnitudes of the width W of the wooden molds were 50.8 mm, 101.5 mm, 152.4 mm, and 203.2 mm. These values of W yielded W/B ratios of 0.5, 1.0, 1.5, and 2.0, respectively.

Once the height of clay compaction was reached, the desired wooden mold was centrally placed in the test box on the compacted clay layer. The clay was then compacted in the test box to the desired unit weight in lifts of 50.8 mm to 76.2 mm. Following the clay compaction, the inserted wooden mold was carefully removed, thereby creating a trench. Measured amounts of sand were then placed in the trench and compacted in lifts of 50.8 mm by a flat bottomed rammer. The model foundation was then placed on the granular trench (i.e.,

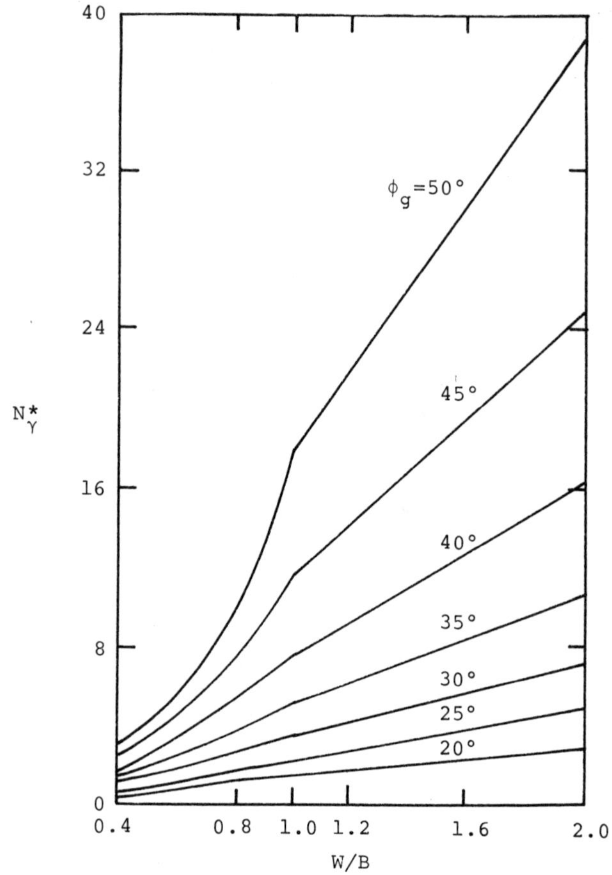

Fig. 5. Variation of N^*_γ with ϕ_g and W/B based on the solution of Madhav and Vitkar

D_f=0; surface foundation condition), and the bearing capacity test was conducted. At the end of each test, some surface cracks were observed close to the foundation. The sand in the trench was carefully removed, and the post-failure shape of the trench was observed. In all cases, a lateral bulging of the trench was evident. The bulging was clearly noticeable at depths of about 0.5B to B.

The sequence of the model tests, unit weight of compacted soil, and the shear strength parameters of the soil are

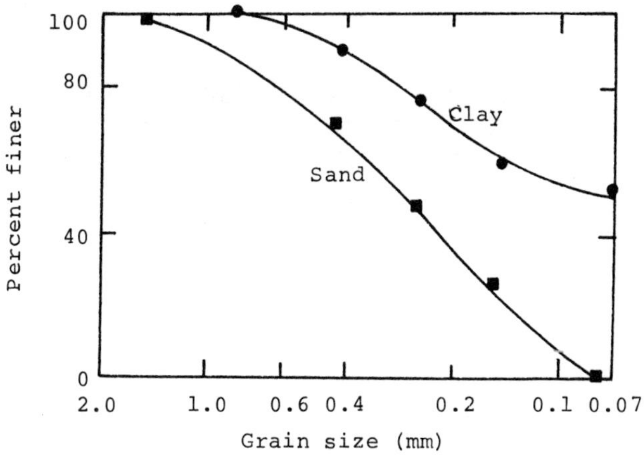

Fig. 6. Grain-size distribution of the sand and clay used for the model tests

shown in Table 1. The undrained shear strength of the clay as given in Table 1 was determined by unconsolidated undrained triaxial tests. The specimens were collected by pushing thin wall sampling tubes into the compacted clay layer. The scattering of the value of c_u was less than $\pm 8\%$ of the average value.

It needs to be pointed out that the distance between the edge of the model foundation and the edge of the model test box in the length side was about 4B for all tests. So, some edge effect might have existed during the tests.

Model Test Results (Tests on Sand and Clay Only)

Figure 7 shows the plot of load per unit area (q) on the model foundation and the corresponding settlement for Test Series I (on clay only) and II (on sand only). A general shear type of failure for the tests on sand was observed showing a peak value. For tests on clay only, the nature of the plots of q vs. foundation settlement was typical of local shear type of failure in soil. The ultimate load per unit area failure was determined to be the point at which the load-settlement plot became practically linear as recommended by Vesic (10). For surface foundation conditions ($D_f=0$)

$$q_u = c_u N_c \qquad \text{(in clay with } \phi=0) \tag{4}$$

Table 1. Sequence of Model Tests and Soil Parameters

Test series	Test Nos.	Test description	Soil parameters
I	1	On clay only; $D_f=0$	$\gamma_c=19.97$ kN/m^3; Average $c_u=10.06$ kN/m^2; Average moisture content=23.34%; Saturation=98.5%
II	2	On sand only; $D_f=0$	$\gamma_g=16.43$ kN/m^3; $\phi_g=40°$
III	3-12	On clay with granular trench; $D_f=0$; W/B=0.5; H/B=0.5 to 5	Trench Sand: Same as Test Series II Clay: Same as Test Series I
IV	13-20	On clay with granular trench; $D_f=0$; W/B=1.0; H/B=0.5 to 4	Trench Sand: Same as Test Series II Clay: Same as Test Series I
V	21-28	On clay with granular trench; $D_f=0$; W/B=1.5; H/B=0.5 to 4	Trench sand: Same as Test Series II Clay: Same as Test Series I
VI	29-36	On clay with granular trench; $D_f=0$; W/B=2; H/B=0.5 to 4	Trench sand: Same as Test Series II Clay: Same as Test Series I

Note: ϕ_g of sand determined from direct shear tests; c_u= undrained shear strength of clay determined from unconsolidated undrained triaxial tests; H=depth of trench; D_f=depth of foundation

$$q_u = \frac{1}{2}\gamma_g B N_\gamma \qquad \text{(sand)} \qquad (5)$$

in which N_c and N_γ=bearing capacity factors. Using proper

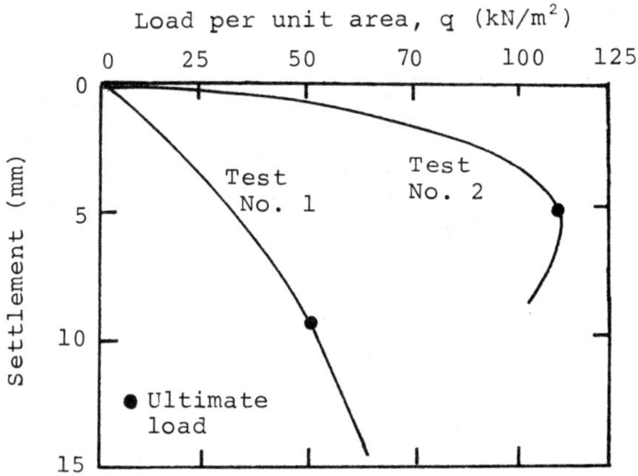

Fig. 7. Plot of load per unit area vs. settlement for Test Series I and II

experimental values of q_u in Eq. (4) yield a value of N_c= 5.25. Similarly, the experimental values for tests in sand, when substituted into Eq. (5), yield N_γ=128. The experimental value of N_c=5.25 is in fairly good agreement with the theory of Prandtl (9), in which N_c=5.14. The experimental value of N_γ=128 is slightly higher than that provided theoretically by Caquot and Kerisel (2) and Vesic (10) for ϕ_g= 40°. However it is somewhat less than that determined by Chen (3) using the limit analysis theorem.

Tests With Granular Trench

The nature of the load-settlement diagrams obtained for model tests with a granular trench (Tests No. 3-36) were primarily of local shear type (similar to Test No. 1). The ultimate bearing capacity for these tests was determined in a similar manner as recommended by Vesic (10). The nature of variation of q_u with H/B and W/B as obtained from these tests is shown in Fig. 8. From these plots, the following general observations can be made:

1. For given values of ϕ_g, c_u, and W/B, the magnitude of the ultimate bearing capacity increases with H/B almost linearly up to a maximum value and remains practically constant thereafter. This maximum value is being referred to hereafter as $q_{u(max)}$.

2. The magnitude of $q_{u(max)}$ increases with the increase of W/B within the limits of the present tests.

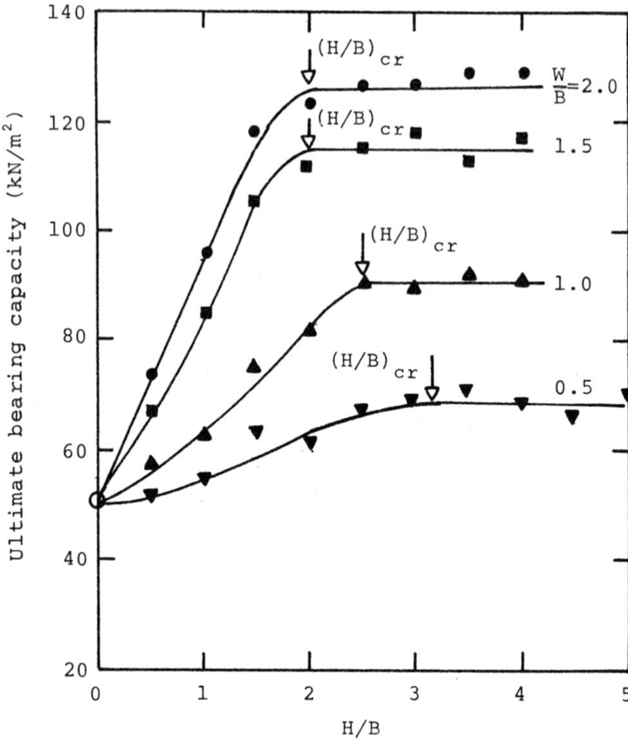

Fig. 8. Variation of the ultimate bearing capacity with W/B and H/B

3. The magnitude of the critical value of $H/B=(H/B)_{cr}$ at which $q_{u(max)}$ is obtained does decrease with the increase of W/B. This is shown in Fig. 9. However, for $W/B \leq 1.5$, the critical H/B ratio is about 2.

Figure 10 shows the variation of $q_{u(max)}$ with W/B as obtained from Fig. 8. For comparison purposes, the theoretical values of $q_{u(max)}$ obtained from Eq. (3) and Figs. 3, 4, and 5 have also been calculated and are plotted in Fig. 10. Based on this comparison, it appears that the theoretical values overestimate the experimental values. The difference between $[q_{u(max)\text{-theory}} - q_{u(max)\text{-experimental}}]/q_{u(max)\text{-theory}}$ varies from about 26% to about 40%. The maximum difference is at W/B=1. The main reason for this large difference is probably due to the fact that the failure in soil at ultimate load is due to bulging of the granular trench. However the basic assumption in deriving the bearing capacity factors given in Figs. 3, 4, and 5 is a general shear

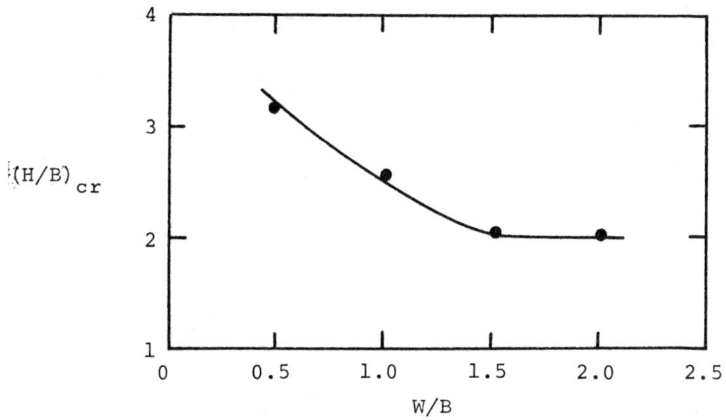

Fig. 9. Variation of $(H/B)_{cr}$ with W/B

failure in soil.

The range of settlement of the model foundation at ultimate load for various values of H/B and W/B is shown in Fig. 11. For a given value of W/B, the settlement level decreased with H/B. For $H/B \geq$ about 2.5, the foundation settlement at ultimate load remained practically constant. This level of settlement was close to that obtained from the bearing capacity test on sand only (Test No. 2).

Conclusions

A number of laboratory model tests for the ultimate bearing capacity of a continuous shallow foundation supported by a granular trench made in a weak clay have been presented. The model test results have been compared with the bearing capacity theory presented by Madhav and Vitkar (7). Based on the present tests, the following general conclusions can be drawn:

1. For a given W/B ratio, the ultimate bearing capacity of a foundation increases with the depth of the trench (H) up to a maximum value [$q_{u(max)}$] and remains constant thereafter.

2. The critical value of H/B at which $q_{u(max)}$ occurs is about 3 for W/B=0.5. For $W/B \geq 1.5$, the critical value of H/B is about 2.

3. For a given value of H/B, the magnitude of the ultimate

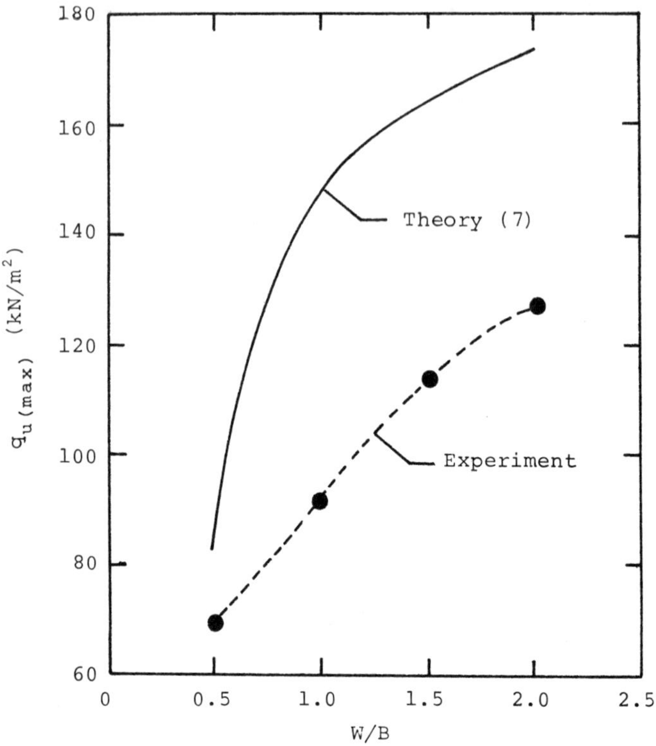

Fig. 10. Variation of $q_{u(max)}$ with W/B--comparison of theory with experiment

bearing capacity increases with the increase of W/B.

4. The magnitude of the experimental $q_{u(max)}$ for a given value of W/B is lower than that predicted by the theory presented by Madhav and Vitkar (7). The largest percentage of difference between the theory and the experiment is obtained at W/B=1. The primary reason for the disagreement between the theory and the experiment is due to assumption of the failure mode in the soil located below the foundation. For the model test results presented in this paper, failure occurred by bulging of the granular trench.

5. The results reported herein are based on small-scale model tests. Due to possible scale effect, the magnitude of the ultimate bearing capacity as determined from the laboratory tests may be somewhat higher than

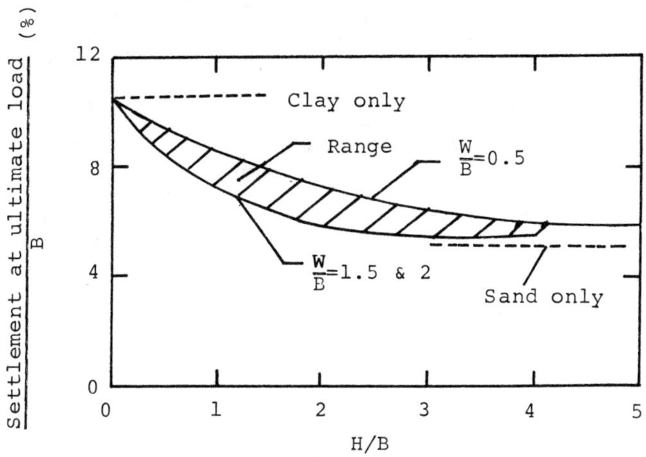

Fig. 11. Range of variation of settlement at ultimate load with H/B

those obtained from large-scale field tests. In such cases, the difference between the theoretical and experimental values of q_u will be greater.

References

1. Barksdale, R.D., and Bachus, R.C., "Design and Construction of Stone Columns," FHA Report No. FHWA/RD-83/026, Vol. I, 1983.
2. Caquot, A., and Kerisel, J., "Sur le Terme de Surface Dans le Clacul des Fondations en Milieu Pulverulent," Proceedings, III International Conference on Soil Mechanics and Foundation Engineering, Vol. I, 1953, pp. 336-337.
3. Chen, W.F., "Soil Mechanics and Theorems of Limit Analysis," Journal of the Soil Mechanics and Foundations Division, ASCE, Vol. 95, No. SM2, 1969, pp. 493-518.
4. Drucker, D.C., and Prager, W., "Soil Mechanics and Plastic Analysis of Limit Design," Quarterly of Applied Mathematics, Vol. 10, 1952, pp. 157-165.
5. Hughes, J.M.O., and Wither, N.J., "Reinforcing of Soft Cohesive Soil with Stone Columns," Ground Engineering, Vol. 7, 1974, pp. 42-49.
6. Hughes, J.M.O., Withers, N.J., and Greenwood, D.A., "A Field Trial of Reinforcing Effects of Stone Columns in Soil," Geotechnique, Vol. 25, No. 1, 1975, pp. 31-34.

7. Madhav, M.R. and Vitkar, P.P., "Strip Footing on Weak Clay Stabilized with a Granular Trench of Pile," Canadian Geotechnical Journal, Vol. 15, No. 4, 1978, pp. 605-609.
8. Mitchell, J.K., and Huber, T.R., "Performance of a Stone Column Foundation," Journal of Geotechnical Engineering, ASCE, Vol. 111, No. GT2, 1985, pp. 205-223.
9. Prandtl, L., "Uber die Eindringungsfestigkeit (Harte) Plastischer Baustoffe und die Festigkeit von Schneiden," Zeitschrift fur Angewandte Mathematik und Mechanik, Vol. 1, No. 1, Basel, Switzerland, 1921, pp. 15-20.
10. Vesic, A.S., "Analysis of Ultimate Loads on Shallow Foundations," Journal of the Soil Mechanics and Foundations Division, ASCE, Vol. 99, No. SM1, 1973, pp. 45-73.

SUBJECT INDEX

Page number refers to first page of paper.

AUTHOR INDEX

Page number refers to first page of paper.